LEAN
MAINTENANCE

JOEL LEVITT

Industrial Press
New York

Library of Congress Cataloging-in-Publication Data
Levitt, Joel, 1952-
   Lean maintenance / Joel Levitt. – 1st ed.
      p. cm.
   Includes bibliographical references.
   ISBN 978-0-8311-3352-8 (hard cover)
1. Plant maintenance—Management. I. Title.
   TS192.L4687 2008
   658.2'02–dc22
   2008010930

Industrial Press, Inc.
32 Haviland Street, Unit 2C
South Norwalk, CT 06854

Sponsoring Editor: John Carleo
Copyeditor: Bob Green
Interior Text and Cover Design: Janet Romano

10 9 8 7 6 5 4 3

TABLE OF CONTENTS

Table of Contents

DEDICATION

The author has had the profound privilege over the last 22 years to work with maintenance individuals and their organizations that fight to do the right thing for their employees, customers, communities, and the environment. While no one knows what they do but themselves, they do the right thing, even in the middle of the night. Even when others in the company are yelling to cut corners, these men and women take the heat and do the right thing.

It is to these people that this work is dedicated. The prayer might be, may we all measure up to their integrity and aspire to attain their stature. The damnedest thing is that they don't even know how important they are and no one thanks them. Well, thanks to all of them and thanks to you if you are one of them.

Our guiding light is Benjamin Franklin. He was the first American to make a living from thinking about and publishing Lean ideas.

Joel Levitt
2008

FOREWORD

I don't think anyone would deny that humanity has gotten the world into a pickle. As maintenance professionals, the question is, what contributions can we make to do our part to change things for the better? The question of this work is what is the contribution of maintenance to the world? That is a big and sometimes uncomfortable question.

Can better maintenance practices save the planet? Improving maintenance practices worldwide is not enough, so probably the answer is no. But it is unforgivable for any group to not do its part. So, while we cannot solve the problems of global warming, or resource depletion, or any of the other problems ourselves, we will do our part. We can also provide leadership in our organizations for the changes that will work toward solving these problems.

Our part is to make our organizations more efficient in all measures. Every breakdown, every part used, even every unnecessary hour we spend, wastes resources. Although the goal of Lean Maintenance is to save money by cutting costs, this objective is secondary to the bigger game of making our companies and other organizations more responsible in how they conduct themselves.

Our responsibility is to use the fewest resources possible to get the product or service out the door. Being responsible is to leave the environment wherever your facility is, better than how you found it. Acting responsible is to be good and protective toward your neighbors and employees. Responsibility is to have standards and ethics that you follow, even if no one is looking. Finally, we want to be proud of how our organizations and leaders conduct themselves.

INTRODUCTION TO LEAN MAINTENANCE

This book is dedicated to the Game of making your product with as few inputs (of all types) as possible, which is part of a bigger game to make your industry sustainable and your organization responsible in the new world of limited resources.

Dedication to fewer inputs is a similarity between this approach to Lean maintenance and other, older approaches. The difference is the reason or driver for the effort. Traditionally the driver is higher profit. Our endeavor is to reduce the use of all resources used to make a product or provide a service. The end result might be the same, but the intention is very different. Higher profits and lower costs of goods sold are the gravy from this process. The meat is being able to produce products with lower and lower levels of inputs, which means consuming fewer resources.

Industry all over the world is getting the message. We want it all! We want wood to build houses and we want forests to visit. We want coal for power and we want clean air; we want low carbon emissions and we want recreational areas after the coal is gone.

In the South African gold fields, a major gold mine is reprocessing their tailings piles because new processes have been developed that can extract gold from the discards of only 10 years ago. Think of the energy and labor savings of not having to mine the ore, or even not having to carry it up the 8700-foot shaft.

A Pacific Northwest saw mill that makes dimensional lumber is getting better yields. After the tree is debarked it is scanned by a laser. A computer calculates the maximum yield from each tree given its shape, size, and length. Yields are up, and we have to cut down fewer trees to provide the same amount of lumber. They even accept the lumber brought down in storms and process it into good 2 X 4s.

Not far from the saw mill is the center of French fry production in the US. A typical plant might process a million and a half pounds of spuds a day. Their imaging systems can see a bad fry as it whizzes by at 30 miles an hour. The bad section gets punched out and fed to the (very happy) local cattle. The French fry producers have managed to use fewer potatoes to make better fries at lower costs. We want it all.

Has maintenance kept up with these improvements? Is maintenance producing the equivalent outcomes by mining the tailings pile, getting out more useful products per tree, or eliminating, even small, imperfections? I think not. I think that maintenance has lagged behind manufacturing in contributing to the efficient and Lean enterprise.

History of Lean maintenance

Lean maintenance was first distinguished as a unique program by observers of the Toyota Production System (TPS) in the early 1980s. The phrase Lean Manufacturing was never used by Toyota but was coined by James Womack in his ground-breaking 1990 book titled *The Machine That Changed the World*.

TPS concepts include: waste elimination (Lean Manufacturing), standardized work practices, just-in-time manufacturing scheduling, and a focus on quality. Shigeo Shingo is generally credited with originating lean manufacturing as part of TPS. He was a brilliant observer of manufacturing processes, and could see the waste in almost every process. Lean was named from that time, but the philosophy has been around for hundreds or even thousands of years.

Womack and Jones in their 2005 book *Lean Solutions* defined lean thinking. The two men were discussing manufacturing a product, but the conclusion completely applies to maintenance. It is said that the process should provide for something actually desired by the customer—in other words, a product that the customer wants or needs. In our business, the products are the services that repair breakdowns, efforts that provide reliability and uptime, and anything we do to assure consistent quality output.

In an interview, Geoff Green, an expert and facilitator of Lean manufacturing practices for SIRF Round Table in Australia, and Joanne Law, marketing head of the Lean Roundtable, described some of the history and attitudes important in Lean approaches. First Geoff

Introduction to Lean Maintenance

explained that Lean came from automobile assembly plants where machines are small, operators are extremely well paid, and high-end (skill-intensive) maintenance may be outsourced anyway. This history explained why all the programs that evolved from TPS were operator-oriented and not maintenance-oriented (such as TPM).

Green and Law went on to explain that the tools they use for Lean are just tools and they are not 'it'; the tools are not the program. What is 'It' is to have the employees engaged in their jobs, aware of the process around them, and concerned with waste. Employees can expose problems and have the power to resolve the problems. Lean maintenance would have the goal of having the maintenance workers be conscious of and concerned about waste.

One of the powerful exercises commonly used to show where waste exists in Lean manufacturing is to draw a circle on the floor from which the Lean team members can see the operation, and park them there for a few hours so that they can look at what is going on, ask themselves questions, and begin to see waste. This same technique can be applied by drawing the circle anywhere that maintenance people congregate (stores, tool crib, or the maintenance shop).

It is stressed that Lean is a tactical tool not a strategic tool. It is not designed for long-term change but rather toward immediate waste elimination. There is a tendency to push Lean toward a strategy for long term review of the maintenance process. There is a question to answer: What would my customer not be willing to pay for, if they knew about it?

In any program, or in any part of life, there are people who go too far and include everything under the banner of Lean (or in whatever else their passion is centered). Green and Law's last comment in the interview was to beware of the lean fundamentalist.

What is it?

"Well what is lean?" you might ask. Lean is what people have always done. If you owned a factory you would practice Lean. In your household, don't you try to run a lean operation? Aren't you always telling the kids to turn the lights off, and to not use so much water?

Introduction to Lean Maintenance

Shingo's definition of lean is an all-out war against waste from both manufacturing efficiencies and under-utilization of people. That is a pretty good starting point. Let's get more specific.

Lean Maintenance: Is defined as delivery of maintenance services to customers with as little waste as possible, or producing a desirable maintenance outcome with the fewest inputs possible. In this discussion we will investigate ways of providing excellent services while minimizing the 10 inputs:

1. Labor (any kind including labor from the operator, mechanic, clerk, staff, and contractor)
2. Management effort (reduce headaches, or non-standard conditions requiring special management inputs)
3. Maintenance parts, materials, supplies
4. Contractors
5. Equipment rental
6. Service contracts of all types
7. Raw materials
8. Energy
9. Capital
10. Overhead

And/or ways of maximizing the outputs:

1. Improved reliability (uptime)
2. Improved output quantity
3. Improved repeatability of process (less variation)
4. Improved safety for the employees, the public, and the environment

The challenge is to produce these lean outcomes while maintaining a long-term, safe, environment, and conforming to governmental statutes and company policies.

IFS is a consultancy in the related area of Lean Manufacturing. They make some promises in their White Paper on Lean Manufacturing (to be found at their web site www.Ifsworld.com). The savings are likely to be similar to the ones that can be found in Lean Maintenance.

Introduction to Lean Maintenance

- A 90% reduction in lead time (cycle time)
- A 50% increase in productivity
- An 80% reduction in work-in-process inventory
- An 80% improvement in quality
- A 75% reduction in space utilization

In formal definitions, Lean is defined as the elimination of everything in the value stream that does not provide added value to the customer or to the product. Another way to approach this subject is to ask the question introduced above: "If the customer knew about this, would he/she be willing to pay for it?" So the customer might be willing for an effective permitting system to promote safety, but might also be unwilling to pay for the hours of waiting for the operator. Anything not contributing value is waste. So waste-free is the same as Lean. Of course waste has many forms that we will be looking at throughout this work.

Lean has four dimensions, or operates in four zones:

Practices: Lean Maintenance can be thought of as a set of practices and attitudes toward maintenance. These practices (like never sending someone to repair something without a plan, bill of material, and tool list) are like exercising, eating right, or self improvement. The improvement from the practice comes from its application over a long period of time. The good Lean Maintenance practices will carry you along, with gradual improvements over several years.

Attitude: Having the right attitude is related to practices. We are trying to produce our product or service with the least input possible because it makes sense from a profit motive and because it is the right thing to do for our environment and for our world. Like someone exercising, even if your attitude flags (I don't want to exercise today), the habit of the practice will carry you over (you don't have to like it—just do it!)

Technology: Breakthroughs in technology can happen at any time. Innovation is either continuous (incremental improvement) or discontinuous (giant leap in improvement involving shift to new approaches). Maintenance professionals who want Lean Maintenance will have to be in a constant search for new technology. The key to using technology is to wait for others (who like living on the bleeding edge of technology) to do the initial "beta" testing and to have a program in place to try some of the successes. Testing new technology in a scientific way (that is with rigorous testing and a control) is essential to know if a new technology is indeed better than the old.

Duration: Savings (of resources or money) from Lean projects are like little rivulets of water flowing back to the company. Over time and with ongoing attention they gather together into massive streams and then rivers of savings. The key is time or duration. One project will not make much difference, but dozens of projects over a few years can make a substantial, quantifiable difference in the picture, as shown by the diagram below.

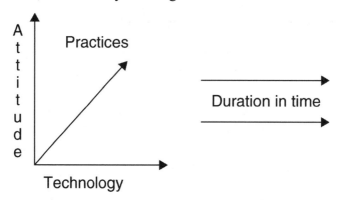

Levels to Lean: Lean Maintenance can be seen in levels. Each level includes the levels below. Each level expresses a higher level of knowledge, investment, and commitment.

Level one—Looks lean superficially: Lean Maintenance efforts start with 'Looks Lean' projects. It is likely that there are easy savings or what is called 'Low Hanging Fruit'. You could just reach out and grab the savings. Savings of this type include re-lamping with compact florescent lamps, or plugging steam and compressed-air leaks.

Introduction to Lean Maintenance

There are great, quick, and low-cost actions that can be taken in this domain. Many organizations are satisfied to operate in level one, and never go to the higher levels.

Level two—Studies to make lean: Once the 'Low Hanging Fruit' is picked you will have to conduct studies and do some research into Lean practices and products. The effort and knowledge required for success is higher. Savings potentials here might include high-efficiency motors, more-efficient starters, or more-slippery lubricants. Improved inspection technology such as infrared cameras also fit in here.

Level three—Lean potential limit given existing technology: At the top of the totem pole is the investigation and adoption of new, different, and more-efficient manufacturing processes. These processes might offer better yields, higher quality, increased up time, lower energy usage, and fewer maintenance-intensive processes. Included would be maintenance process re-engineering, using some kind of hybrid or fuel cell technology, integrating technologies to make all systems work together, and major changes to new technologies for efficiency and improvement.

The nature of Fat Maintenance

There are important questions that could help big organization attack their tendency to get fat. Why is it that people, institutions, and especially big institutions, always seem to go toward fat (with few exceptions) as they age? Is there any relationship between human middle age and organization middle age? Is this preordained? Is this entropy (the tendency for systems to go toward disorder)? These are particularly important philosophical questions. If there is any way around this stage it would be important to know.

Another useful inquiry: is the goal of Lean Maintenance to be totally lean or is a little fat ok. Is a basketball court or gym in the plant against Lean? If a company has the resources, can it put in day care for the employees' kids or build an outdoor picnic area? Can some fat be good?

Introduction to Lean Maintenance

The issue is how we define fat. It is perfectly OK to have a gym because it does not impact the leanness of the basic operation (except using a few hundred square feet). In our definition, giving a little back to the employees for their health or enjoyment is not fat at all. It is by removing the fat that resources are made available for return to the employees and shareholders. Of course, the Lean fundamentalists disagree, saying that any square footage not dedicated to making product is fat. So a garden with picnic tables, or a daycare center, is fat for them but not for me.

So, what is fatness? At a survival level, fat is stored resources. Fat is stock stored against a famine. At emotional levels fat is satisfaction of phantom hunger. At a sensory level, fat is enjoyment of the fine things in life at levels and in amounts well above sustenance levels. The fat in organizations is all this and the inflexibility, lack of insight and feeling of struggle to get things done that goes with it.

Many great companies that were known in the last few generations as Lean have become bloated. In companies, fatness may come about when, for that time and place, survival and competition have been handled. However, survival and competition are never completely handled. but they can be handled for a time and in a place. The practices and attitudes that developed during the growth phase still continue to work reasonably well in the fat phase. Once the outside facts change enough (the marketplace, the technology), the company can fail, or be forced to reinvent itself to survive.

The best company example of this life cycle is IBM, once the pre-eminent computer manufacturer in the world with a market share over 70%. There was the 'IBM way,' which worked well for decades and set the stage for today's computer culture. Then the computer world changed.

Companies introduced small personal computers. In fact, IBM was an accelerator of the change with the introduction of their IBM PC. That introduction itself was a break from the past because it used standardized instead of (its usual) proprietary parts. The short version of the story was that IBM's fortunes fell dramatically and the company encountered hard times (for IBM). The company looked fat and without the ability to reach to the marketplace.

Introduction to Lean Maintenance

Just before the Internet boom, IBM had successfully reinvented itself as global integrators, consultants, and middleware suppliers (software that sits between the operating system and the user applications). The company is now re-invigorated and it looks entirely different from how it did in 1980. It is now a faster, leaner, and extremely effective force in the computer field for large businesses.

We associate fat with laziness. Fat organizations tend to be bureaucratic and inflexible. IBM filled this definition to a 'T' before its re-invention. In these organizations it is hard to get stuff done. The focus is more internal (rules, procedures, power) than external (toward customers and market needs). Lean organizations tend to be in action, and have less baggage to carry around, so getting things done is the expression of the company.

Consider the stories of John D Rockefeller and Standard Oil. In the early days, Rockefeller's Standard Oil was known as a Lean producer. Its rapid growth was not all due to his abuses of size and power that caused the government to break up the company. Initially he paid enormous attention to the cost and consequence of every decision. In one story he counted the number of drops of solder it took to seal a barrel. He realized he could get by with fewer drops. Now that is Lean!

Once Rockefeller became powerful and began to use his power to force his competition out of business, history has said, his behavior was unacceptable. He broke the law. "In 1911, the Supreme Court upheld the lower court judgment, and forced Standard Oil to separate into thirty-four companies" (Wikipedia). These companies became the leading oil companies in the world including Exxon-Mobile, Conoco-Phillips, Amoco and Sohio (now part of BP) and others.

The important conclusion is how the successors are oriented toward efficiency and Lean. In fact, none of the 34 companies would be known now as a Lean producer. So from the Lean roots and John D. Rockefeller we have inherited some very fat companies. Their Lean roots aren't even part of their advertising or identity.

No such thing as Lean

Almost every organization that watched its pennies while they were growing now has trouble in watching its millions. Part of the issue is that there is no such thing as a Lean company. What does this mean? There is no magical place called Lean where you can arrive. There is only a Lean company here and now. More properly, companies move toward or away from Lean operations every day.

There is no permanence, just process and direction. Any company that considers itself lean (just like any company that thinks it is world class), is engaged in self deception. In a moment the leaders in the industry will fall and the followers will lead (or people who just entered the industry and didn't exist a few years ago will lead everyone). Complacency is the enemy of a truly Lean enterprise. Simple changes in technology tomorrow can open avenues not contemplated by any expert today.

There are hundreds of examples of Fat maintenance. For example, organizations can have too much PM. We remember the words of John Wanamaker (a Philadelphia, PA department store founder in the early 20th century) when he said that he knew that half his advertising was wasted. He said his problem was that he didn't know which half.

Half of a good PM effort is wasted. Nine tenths of a bad effort is wasted. One thing to consider is, what if you have a machine with a low consequence of failure? Let's also say the PM cost exceeds the cost of the avoided failure. Is it Fat to PM it, and is it Lean to let it run to failure? For some people, choosing a breakdown strategy would be a scary thought (though it might be the Leanest choice).

Lean is both a continuous and a discontinuous process in that you can develop and improve your leanness by doing training and projects. Lean is discontinuous because as lean as your racks of relays are, they might be the height of fat in comparison to the solid-state world.

There's a series of steps that make any service or product. Each step adds a little value to the product. This sequence is called the value stream, and it is comprised of all the steps from the identification of the need to the satisfaction of the need. In maintenance, the stream includes the work request or notification, the work order, job planning, coordination, scheduling, execution, and communications. All the subsidiary streams, such as material management, are value streams that flow into the main stream.

Introduction to Lean Maintenance

Within the value stream there's a flow from step to step. Our goal is to provide maintenance services without waste or waiting. We want to eliminate both wasted execution and waiting time, and all excess materials. The waiting includes time expended by both the customer and the tradesperson. Let the customer pull value from the provider; customers say when they want the product and what product they want. In the maintenance world, customers define the uptime needed and the quality requirement. Then they schedule windows where maintenance can be provided.

Ideally, lean manufacturing seeks perfection in providing the product with no waste and full value for the manufacturing investment, thus defining lean manufacturing. This definition will be shifted here to suit maintenance.

What are some of the barriers to Lean?

There is resistance to Lean Maintenance. This resistance can be seen in any of the four dimensions (Practices, Attitudes, Technology, and Duration).

1. Reluctance to adopt new technology to solve old problems with known solutions (such as replacing incandescent lamps with compact fluorescent lamps).
2. Concentrating on initial costs and not realizing the cost or time impacts of the waste (such as extended drain intervals of newer oils).
3. Being ignorant in areas where an expert would know of a better product or process (using stabilizer in swimming pools to reduce the use of chlorine).
4. Being complacent.
5. Not being concerned with the efficiency of the process or with the waste produced (particularly when profit margins are high).
6. Resistance when producing waste is cheaper than re-use, re-purpose, or re-cycle.
7. Having all the attention on other targets such as production level, quality, or safety.
8. Having an aesthetic that requires waste (such as an architect wanting a vaulting atrium with glass roof in Arizona).

9. Having an aesthetic that requires obscure products (like lamps that emit certain colors not available in florescent) and not accepting something close.
10. Fun (like having a 300-HP car rather than a 200-HP car).
11. Waste requires less attention and it may be easier to manage than Lean.
12. Unusual quality requirements where useable, almost-good, components are scrapped.
13. Where consumption makes a statement (such as the lagoon in the middle of the desert at the company headquarters).
14. Marketing might drive waste by super-sizing products, which people then throw away in larger amounts.
15. Lack of production planning (like the bakery that over-produces and then has excessive leftovers).

Problem

There is a problem with the notion of Lean Maintenance. The problem is that, if you follow Lean Maintenance through to its logical conclusion, Lean Maintenance really means no maintenance. Or rather, it means no need for maintenance. It means a factory with equipment that doesn't need to be maintained. Wouldn't that be the Leanest Maintenance of all? If we could build a manufacturing process that would reliably make a product that didn't require any maintenance, wouldn't that be interesting to top management?

After all, maintenance itself (unless you do some interesting manipulations with semantics and logic) does not directly add value to a product. This value adding proposition is part of the definition of lean maintenance. So, we have a little bit of a dilemma.

The solution to this dilemma is that as yet there are no truly maintenance-free factories. Until there are, we have to make maintenance lean. When there are maintenance-free designs, we will have to reconsider this position. So, on the way to having maintenance-free systems, you can take on the fat in maintenance.

Some of you may watch *Star Trek* (one of my favorite series that I think has captured many issues of maintenance management over time). I tell people that if you ever want to know something about maintenance, just watch *Star Trek* and see how maintenance issues are treated. Everything you need to know about maintenance has been in a *Star Trek* episode.

In the first *Star Trek* series, back in the late 60s, Scotty was the chief engineer and Scotty would fix things. Scotty would get down underneath this thing and he would wriggle around some complicated-looking tool and he would fix things. He'd actually get dirty.

And in the second series *Star Trek, the Next Generation,* Geordi La Forge was the chief engineer. In a hundred episodes or so, Geordi rarely fixed anything. He would have a problem (there was almost one in every episode); he'd walk over to a computer console and would reconfigure the warp couplings or something like that, by punching a bunch of weird-looking symbols.

And then in *Voyager,* it's even better. Participants in the *Voyager* series didn't even discuss breakdowns because the ship was biological and fixed itself. B'Elanna Troy was the chief engineer and her whole mission was getting more power. She spent her time getting more power out of the warp engine (the reason why is beyond the scope of this discussion).

And these items illustrate exactly what are the three recent generations of maintenance. The first generation, in the mid 1960s, was the super fix-it guy who could fix anything. He was an older person (compared to the rest of the crew) but up-to-date with all the technology. The second generation in the 1990s was the young, smart, computer guy. And in the third generation in the 2000s, the ship was biological, and could just grow new circuits.

Machines that can fix themselves are what your bosses dream about, by the way.

These are battles that have been fought for years

Many of us have been fighting these lean battles for years. In a recent training session on Planning in an ore processing facility, I was amazed that, without prompting about waste and the value stream of maintenance, or any preparation to set up the discussion, the maintenance workers and plant operators came up with a great collection of ideas about where the waste was located.

Introduction to Lean Maintenance

You can see that waste is foremost in the workers' minds because unlike managers, the waste is in the workers' faces. Logically, the list of ideas can be divided into a few areas. These few areas could be used to start the lean discussions within the plant. If you embark on a Lean journey, always ask yourself the question: where is the lowest-hanging fruit in each of the areas? Nothing beats having success on your first few projects.

Waste discussion from Alcoa, Point Comfort, Texas July 2007,

- The biggest waste is searching for and ordering parts. It seems wrong parts are constantly ordered or received. The whole process around parts is a huge waste of time.
- Getting refurbished components from outside vendors with incorrect specifications. And inadequate follow-up on outsourced rebuilding.
- Parts that are pulled for a job and not used are thrown away because it is too much trouble to return them.
- Long-lead-time parts are not set up on the system or stocked (the plant is 50 years old).
- Time wasted going to jobs unprepared. The maintenance people frequently don't have the materials or tools, and then have to leave the jobs to collect the stuff.
- Planners do not have job knowledge or know the job requirements.
- Starting jobs before you have all the parts and equipment (this is a daily problem).
- Showing up on jobs and not being able to get access to equipment because it's not locked out, cleaned, or ready for maintenance.
- Not having the right tools in the tool room and having to improvise.
- When you get tools from the crib, they frequently don't work.
- Lifting gear (cranes, Broderson, etc) break down too often. Mobile equipment takes too long to get back from the garage.
- The biggest waste is sharing large equipment. When ever a mobile crane is needed it seems like someone else has it. It then becomes necessary to steal one from another unit.

Introduction to Lean Maintenance

- Craftspeople do not get ready to use the crane when they order it (because they know they will not get it right away anyway. When they do get it they might even hide it).
- The issue of cross training of the General Mechanics (GMs). There is a wide variation in skill sets among the GMs. Some can weld plate but not pipe, some can align pumps. These deficiencies cripple the crews, create inefficiency and put a lot of pressure on the GMs with more skills.
- A huge amount of time is wasted in re-inventing the wheel. We have a 50-year old plant, yet we have to invent ways of doing things that have been done successfully before.
- Only patching and not fixing the root cause of the problem due to cost, budget or production concerns wastes a lot of time.
- Poor PMs (not well-directed, personnel skills, often not equal to the task).
- Inflexible break and meal times.
- The Work Requested is incorrectly scoped or defined, leading to the wrong job being done.
- Lack of coherent priorities means workers are pulled off scheduled jobs to work on emergencies, which causes lost time and interferes with the timely completion of PMs (which in turn causes more emergencies). Jobs sometimes come up that are called emergencies but they are not really emergencies.
- Excessive emergencies interrupt scheduled work. Maintenance seldom seems to have a chance to finish what is started in one go.
- Too many managers and not enough workers to do the work. On a valve change, too many supervisors standing around trying to rush the job.
- Supervisors have no idea whether the job is ready.
- Problems with scaffolding. Only contracting for two crews to cover the entire plant so that maintenance frequently has to wait excessive amounts of time. Using a contractor is often a waste of money and causes more problems than it solves.

It might be apparent that some of these items describe waste that is easily accessible or "low hanging fruit" and some items require significant change. As the processes for Lean Maintenance projects are described, you can revisit this list and see even more project opportunities.

Lean Maintenance projects are fun!

The truth is Lean Maintenance is fun. These investigations and projects are going to be the fun stuff, and are the part of maintenance that, if any of you have ever come up through the maintenance ranks, are maintenance engineers, or have hands-on experience designing stuff; this may be the most fun part of what you do.

You'll soon start to see some of the things we're going to address. The exercises in this work can be done with the people in the crafts. This work has been done in a wide range of organizations with a very wide variety of participants.

In a school (described in detail in a later chapter), we worked with the groundskeepers, who didn't understand English. I would give instructions and then one of them would translate it for everyone at the table. And the workers would do these projects, and you could feel the energy. So even though fun is against the grain in maintenance circles, we're going to shoot for that as part of these projects.

Sometimes the negative result is positive

The groundskeepers I just mentioned did a project on edging blades. They felt that a more-expensive edging blade would be Leaner (cheaper per foot of edging and less effort to use). So we bought the expensive cutting blade, but it turned out that their old cheap blade performed about the same as the expensive blade. To compound the problem the cheap one could be sharpened but the expensive one had special steel alloys and a shape that couldn't be sharpened. So it was clearly Leaner and cheaper to use the cheaper blade.

The team felt really bad. They felt bad because their project had failed. I said that I considered it to be a completely successful project, because it gave us information about the process or product that we didn't have before, and we now had a new basis for thinking about things. People were much more conscious of the whole blade issue after that than they were before. So, a good project might be successful even if it has a negative result. The next time a new blade comes out, they can give it a try and test it against their benchmark.

What does waste look like?

Is it non-added value? Actually, that's what waste doesn't look like. What does waste look like? This is not a high-level question; I'm asking a kindergarten question here. Waste looks like garbage, a spill on the floor, extra parts in the scrap bins, personnel standing around.

When you take on Lean you get sensitivity to what waste looks like. You can walk into your place and see waste when you couldn't see it before. Once you start to think this way, you walk through the storeroom and when the dust is really thick on something you start to think of questions. Is the part there for a really good reason (I don't want to minimize that in the maintenance world)?

It may be what we call an insurance-policy spare, which you expressly have on the shelf to not use. But there's stuff in the stock room that you won't use, which is different from "not use".

In the stock room at a large coal-fired power plant they had this large part, it looked like the head of a cylinder. It was covered with a thick coat of grease and dust (so it had been there a while). I asked, "What is that?" The stock room manager answered "Well, I don't really know what it is. And neither do any of my people. And, in fact, I invited the old maintenance guys in here that would have been around when this was used before, and they don't know what it is either." Now that will qualify to me as some kind of waste, although we don't know what kind of waste because we don't know what it is. And, of course, what everybody is afraid of is making a mistake and getting rid of it (and finding it is needed the next day).

An automated assembly operation had a $25,000 power supply for a robot that was never supposed to break. The company bought it because it had a six months lead time. We call that an insurance policy part. It's a well-established strategy that there are certain things that are so onerous to not have, that you have them. **These parts are not waste.** The question is, what about everything else? What about the SKU (stock keeping unit—single part number) that is not an insurance policy spare. That's what we want to attack.

Introduction to Lean Maintenance

Once we start to eliminate waste we start to see what waste looks like. Consider the US Mint. They blank out 11 coins at a time—quarters in this instance. The job is run on a 200-stroke per minute Shuler press. The blanks are cut from the clad metal coil like cookies being cut by a cookie cutter. The blanks (round disks before they are coins) pass through the press and fall on to a conveyer belt. The belt takes the blanks to the coining operation.

US Coins are blanked and coined in separate operations. Under the Shuler press there's a big pile of blanks on the floor. Every blank on the floor cost the Mint $0.06–0.07. The manager said, "Well, what happens is, the blanks hit the conveyer belt and a few roll off." We decided to modify the chute underneath the press to better control the blanks falling on to the conveyer belt. To test the idea we designed an extension to the chute with cardboard and duct tape. We cut up a cardboard box and wrapped it around the chute so that the space was just more than the thickness of the coin.

When the blanks came out on the conveyer belt, any that were standing up were knocked over by the edge of the new cardboard chute. The blanks no longer fell off the belt onto the floor. The number of blanks that fell off dropped to 1 or 2 a shift. The savings was in the thousands of dollars. The impact on maintenance was that the maintenance effort was divided up among more shippable units. An hour a day cleaning up the blanks was also saved.

You have to have an idea where to start to look for Fat. This book will spend several chapters on this issue of where to look. A certain kind of vision is needed that allows you to walk into your plant or any plant and get the tempo of the place.

Look around and see the cleanliness of a "5S" shop (to be discussed in its own chapter). If everything is really tidy you can immediately see problem areas. You can tell when a place is well organized or not well organized, when there's debris under the tables or not. You can see that kind of waste. That is the obvious stuff. There is also wasted effort, time, wasted energy, wasted spares, etc. that you can't easily see.

Introduction to Lean Maintenance

Not every organization should undertake an effort of this type.

This is a tough conversation. If a company is driven by short-term goals exclusively, then Lean Maintenance programs might not be appropriate. It is almost as if the profit-making corporation is designed to be intrinsically against good long-term maintenance practices.

We are not talking about greed here. But we are talking about having a game called business, where one of the rules is to look at the profit numbers for each short interval. If we look only at short term economics, the Future Value (FV) of the maintenance effort discounted to today, never seems to equal the Present Value (PV) of the investment.

This is a pure devil's advocate position. But keep in mind that this view is held by many smart people. In a later chapter we will discuss the leaks in the pipeline in Alaska. As an example, place the savings over 30 years by not doing all the maintenance necessary to avoid the leaks on one side of the equation. On the other side, place the costs to fix the leaks 30 years later. If you do the math, small amounts of money saved over 30 years (this is a mortgage payment type of problem) can justify huge spending on leaks now. Of course these money calculations ignore the impact on the environment, but that impact is not built into the structure of the typical corporation. We have to fight that position with all the tools available. But if the company is dedicated to only short term gains, we may lose the argument.

Given all the kudos given to Lean maintenance it might be a surprise to know it is not always a good idea. Lean Maintenance can be a demotivator unless there are several things that are committed to and present. Some of the attitudes needed for success are:

1. Commitment that efficiencies gained will not result in layoffs.
2. Commitment to follow through on at least 50% or more of the projects if the engineering and economics work (the more projects you follow through on the better). The president of Sony, in its heyday, spent a good deal of effort following

through on ideas from employees, even those of marginal value. He said that every project done was a motivator to the entire workforce and was well worth it, even if it in itself didn't make money.

3. Ability to commit the resources to the effort, including time off from regular duties to work on projects, small amounts of money to purchase experimental materials, and management/staff coaching time.
4. Ability to allow flexibility in purchasing non-standard items from new vendors on a rush basis.
5. Ability to allow people to cross functional and trade lines, and to encourage workers to talk directly to the appropriate experts in the accounting, purchasing, legal, and engineering departments.

CHAPTER

1

Distinguishing Lean from Everything Else

Benjamin Franklin says: Time is money.

Lean Maintenance is mistaken for a whole host of other efforts. In this section we will distinguish Lean Maintenance from some of the other programs that are occasionally called Lean. We will also show how some efforts in the past were really predecessors to Lean Maintenance.

But first we will discuss a program that is Lean. Kaizen is part of Lean. Kaizen is a Japanese word denoting a philosophy of continual improvement. There are two types of Kaizen that apply to Lean Maintenance; Flow Kaizen-value stream improvement, Point Kaizen-Waste elimination. Kaizen events are short term Lean efforts, where teams blitz a shop and make many improvements at the same time. The event is generally organized around a subject. A good source of information on Kaizen is the *Lean Manufacturing Pocket Handbook* mentioned in the bibliography.

Beware of phantom savings

What would be a good waste reduction project? If we want to be lean, we want to reduce the waste, what would constitute a good project? A good project reduces waste that is obvious. The project should save money directly. Now, there's a huge problem here, with projects that save time (labor). If we are saving (just) labor hours, we grapple with something called phantom savings.

Phantom savings drives accountants nuts; in fact it drives everyone who thinks deeply about it nuts, because phantom savings is not traceable to the companies' financial books. For instance, let's say we save 10 minutes in a meeting by better meeting etiquette or by more discipline during the meeting, and there are five people in the meeting, we pick up 50 minutes a day. Where does that savings show up on the books?

The only place that such a savings could conceivably show up is if somebody is fired. So that labor savings becomes a phantom savings, unless the time that's saved can be used on something else that will reduce costs of operation. So you have to just be careful when you're doing these kinds of things. A lot of people might say "Oh, we saved so many hours." And then an accountant might say, "Great, by the way, who is leaving?" They saved half a maintenance welder. So which half of him is going to go? Remember, we already promised that no one goes as a result of the efficiencies from Lean Maintenance projects.

When we're deciding on which projects to do first, we should pick projects with bookable savings. Either we want to have our output (production) go up or our input (materials, energy, labor or overhead) go down, in a measurable way.

If you use contractors for labor you have a great opportunity. Saving contractor hours is a bookable savings (because you can send them home without severance pay). If the plant went from six contractors to five contractors, that's savings that is somewhere on the books.

There are real cost savings and there are phantom savings. Real cost savings flow to the accounting system and appear on the books. Phantom savings appear on reports and can never be tracked to the accounting books.

Some examples of Real savings (Note that not all real savings appear on the maintenance budget. Some are below the so-called water line and accrue to other departments)
Reductions in payroll (personnel)

Distinguishing Lean from Everything Else

Non-replacement of personnel lost through attrition
   because we don't need them anymore
Reduction of overtime
Reduction of billing from contractors
Reductions of material used
Reductions of inventory on shelf
Reduced expenditures for tools and equipment
Reduced equipment rental bills
Reduced demurrage (rental of tanks, rail cars, ships)
Reduction of regulatory fines
Closing a satellite operation with consequent reduction
   of overhead
Reduction of energy usage (large enough to be recognized)
Reduced raw material usage
Reduced number of production machines due to increased
   uptime
Reduced operator personnel resulting from fewer machines

Phantom savings

Reductions in labor without realizing any savings
Small reductions in energy usage (unmeasured and unproven)
Small reduction in production machine usage
Reduced hours of compressor usage due to leaks being
   fixed (unless you can prove electricity savings)
Increased uptime or availability when the product is not
   sold out

For example let's consider a PM that takes 3 hours a month and does not use materials. We decide the PM is too frequent and we reduce the frequency from monthly to quarterly. And let's agree there was no increase in breakdowns or adverse events. Calculations show we "saved" 24 hours a year. Where did the savings go? We say that the time is now available for other valuable maintenance activity. This time is phantom savings.

If we sent home a contractor 3 days a year as a result of this PM frequency improvement, the phantom savings would be realized (translated into real savings). If we could decrease overtime the savings would also be realized. Or if the PM used a $25 belt each month and we dropped the usage from 12 to 4 a year, we could show real savings of $200.

Getting phantom savings are good, but we act as if the real and phantom savings are the same. They are not the same, and should be presented separately. Hard numbers people (accountants) are extremely suspicious of phantom savings. In the real world they never realize those savings. Consider the way your audience listens to talk about savings. There will be significant skepticism. By stressing (a smaller amount of) real, as opposed to (a large amount of) phantom, savings you will be answering the biggest question (which is sometimes not even asked explicitly). Phantom savings are nice to have but not as nice as money in the bank.

People argue that saving time will enable the team member to concentrate on something else during the time saved. This argument may be true. But they also might turn it into a happiness moment, an idea put forward by the Gilbreths (discussed later in this chapter) at the beginning of the 20th century. Lillian and Frank Gilbreth said that, by teaching everyone the one best way to do the job they could get their work done and have time for a happiness moment. It didn't take long for happiness moments to go the way of the dodo bird (to extinction).

Sometimes Lean can look Fat

On a bigger scale, if you read trade magazines you will find articles about savings from adoption of this or that productivity improvement program (new software, a new gadget, or a new way to look at maintenance). The savings are always impressive. Like all phantom savings, rarely if ever are those savings distinguishable when we take a before and after snapshot of the organization's books. If the savings are not visible, the vendors are touting phantom savings.

This statement is not to say that phantom savings are not important, they are. Phantom savings can really be used for important work. It's just that the Return on Investment will show up as a result of only the impacts on the costs, either above or below the water line, and not from the theoretical savings activity. Phantom savings can also accumulate, and when there is enough they can be converted or will naturally convert themselves into real savings. Phantom savings can also be a guide or a pointer to real savings.

The situation of phantom savings could very well be worse than just no impact on the books. Consider the impact of a major effort toward planning and scheduling the maintenance effort. Conservative estimates show productivity could improve by 25%.

Most places don't implement Planning and scheduling and then lay off 25% of their people. Most places have excessive identified work (in their backlog), and use the gain in productivity to accelerate the speed with which they work their way through the backlogged jobs. This strategy results in backlog reduction and timelier customer service. Without a lay-off or reduction in overtime there are no savings in maintenance costs. Eventually, the firm might be able to reduce overtime, contractors, or even headcount through attrition (converting phantom savings to real savings).

Each job takes a shorter time (on the time clock) when materials, tools, permissions, and drawings are available when the job starts. More jobs run smoothly. Then something strange happens. Those additional jobs will consume materials. The up-tick in material usage will be real (not phantom) because more jobs will be run per week. The improving productivity might adversely impact the maintenance materials budget (by using it up faster).

In addition to running through the backlogged jobs, usually there are additional jobs that didn't make it to backlog originally, because no one had confidence that the job would ever get done (this position seems particularly true for infrastructure jobs). Some of these jobs get captured and added to backlog. Eventually, when the backlog is reduced to a manageable level, the whole plant will run better. Fewer corrective jobs will break down waiting for maintenance to get there. Years later there might be an opportunity to allow the maintenance crew to drop in size naturally as people retire and leave.

Is Lean a Religious Issue?

On a flight to the west coast from Philadelphia, I had the great opportunity to sit next to a Minister of a large church. We got to talking and I realized that he knew quite a bit about maintenance.

Chapter 1

As a head of a church, with a building and a school he had faced many maintenance problems over the years. He surprised me when he said that fundamental issues of maintenance and religion were related.

He said that when children are born, or people pass away, or couples get married, there is no problem getting people into church to pray. The problem he said is the decades between these events. It is hard to get people to come into church when nothing of importance is going on. But that time is important, it is time spent building up the spiritual muscle to withstand whatever life has in store for you.

The minister went on to say that, when they had a leaking roof and minor floods after a string of spring storms, he had no problem getting money and congregational attention to fix the roof. But he had no luck getting (less) money in the previous 2 years to fix the roof so it wouldn't leak in the first place.

Many of our companies are just like that. They wait until after a crisis to consider maintenance seriously. One of the battles of the maintenance war is with human nature. We put off paying for maintenance because we either don't believe there will be a problem or we don't really even want to think about the inevitable decline of the assets we are using. We consider this approach Lean. It isn't in the long run.

Competition determines success

Lean maintenance exists by itself in a particular plant. Once you go through the exercises and "Lean up" the process you can look around and declare yourself on the way to a Lean Maintenance operation. But there is a major aspect to Lean that is not self referential. Ultimately the success of the organization might hinge on whether your plant is Leaner than those of your competitors. When there is a major shift in technology or taste, sometimes even a lean plant will be shuttered (such as when there is no longer any demand for the product at any price).

Toyota's Leanness gave them a competitive advantage against first, other Japanese car companies, and then against global competitors. When the cyclic automobile market turned down, Toyota could lower prices to build market share without incurring losses. Leanness made this strategy possible for Toyota. In the US, manufacturers use valuable incentives to sell cars during downturns, and are prepared to lose money to maintain market share. Problems arise when a company is weakened by successive downturns and cutbacks in development.

Walmart distributes products at a lower cost than its competitors. Its Leanness enables it to maintain low prices and still make a profit. In both examples, if history is a good guide, someone will eventually come along with a Leaner strategy, possibly enabled by a new technology. When that happens, the Toyotas or the Walmarts of the world will have to rethink their processes, start new Leanness initiatives, and adopt new technologies.

Efficiency and lean

Is Lean the same as efficiency? Certainly the origins of efficiency and efficiency experts would be in complete accord with what we understand as Lean. The efficiency experts had their heyday after World War II and on into the early 1970s.

These efficiency experts had their roots in the scientific management movement of the turn of the 1900s. Fredrick Taylor is generally credited as the father of the scientific management movement. His approach was to look at the work content of a job, and time everything done with a stop watch, to determine the "one best way," and create a (sometimes) significant productivity improvement. Predecessors to Taylor were the couple of Frank and Lillian Gilbreth.

The Gilbreths studied bricklayers among other trades and activities. Not surprisingly, they found a wide variation in performance and quality. Their studies showed that when the slowest brick layers were taught the techniques of the fastest bricklayers they could almost duplicate the speed and quality of the fast ones. The variation narrowed dramatically. To the Gilbreths, and later to Taylor, this approach showed that there was "one best way" to do any job. The efficiency expert's job was to study the work and determine the one best way.

It was found that variations in the speed of doing maintenance jobs was far more likely to be related to how the job was done than to how fast the mechanic was working. Scientific management techniques, as taught to Industrial Engineers, are still performed today and are enormously useful. The techniques are particularly useful where the same activity is done over and over again. Scientific management task analysis is time consuming and requires significant training to master.

Traditionally, maintenance activity has not lent itself to the how approach because the work is not repetitive and because conditions are frequently so different (rusted bolts, bent housings, etc.). The US government did try to study maintenance activity and produced a series of documents called Engineered Performance Standards or EPS. EPS encompassed all activities that would be needed to maintain a Navy Base (having been sponsored by the Navy). EPS documents are still available.

One company is famous for their attention to the way work is done. UPS has a group of industrial engineers who study maintenance (among all the tasks at UPS) and determine the best way to perform any task. The company then teaches that technique and achieves excellent results, gaining competitive advantages over competitors trying to enter their core business.

Observing work, and training people in the "one best way," is useful (although people are not always open to learning new ways to perform jobs). The focus of Lean is to eliminate the larger pool of non-productive and marginally-productive time that surrounds the productive bit. There is significantly less focus on the "how" of the work, although some study, training, and attention would not necessarily be wasted. In production, frequently, the Lean effort required material bins and tools to be moved, to find the best way to do an operation.

But Lean might not look like what we think, or even what we are led to believe by experts. Lean approaches are built into the cultural traditions of the US. The pilgrims who landed at Plymouth Rock were well known for their thrift and "leanness."

Growing up, one of the parents in my neighborhood owned a factory that made steel shelving. He never spent a cent that he didn't need to. He made trade-offs to run the operation with the least number of inputs and maximum outputs. For example, in whole sections of the plant the lights were turned out if they weren't being used.

Distinguishing Lean from Everything Else

The owner bought only old machines and had the gift of repairing them with objects at hand. Junk parts filled sections of the warehouse, and occasionally yielded critical spares (with some machining or welding).

Despite his small volume, my friend managed to make a profit (enough to send 3 kids to college and provide for a luxurious retirement) for over 30 years. The point is that there is lean by the book and lean in practice. Lean in practice might not look like the shiny, ordered factory we all look up to. It would have taken years (if ever) to provide a return on investment (that could have been millions of dollars) from cleaning up that place and bringing the equipment to like-new condition. And there still would be a question whether all that effort would result in a cheaper and higher-quality product (his product was extremely durable and was the shelving of choice for local companies).

Lean is long term and immediate at the same time. The projects are designed to provide immediate gratification. Generally Lean projects are not multi-year efforts but rather immediate ones. Lean is long term in that the consequences are continued lower costs, increased quality, and improved morale with employee development. How do we know Lean is working? Look at the bottom line. With an effective Lean program in place, costs of maintenance (other things being equal) will gradually drop. You'll see the effect immediately and it will be ongoing for 1 year, 2 years, 5 years, even 10 years.

CHAPTER

2

Is There a Right and Wrong Way to Lean?

Yes, there is a wrong way to approach Lean Maintenance. All the examples from the section on safety and Lean maintenance are wrong approaches to Lean. When the edict comes down from on high "cut 10%..." or "We're Leaning up your plant, reduce head count to 24", that is a wrong approach to Lean Maintenance. If the cuts are not supported by changes in tools, techniques, or approaches, that is wrong. If the cuts do not take the real needs of the equipment and the life cycles of the equipment into account, that is wrong. In fact, with a Lean approach, the savings will flow up to the ledgers instead of cuts flowing down.

The difference is striking. As a parallel example, consider the effect if you as a parent send down an edict to cut household expenses. Let's say overtime has been cut out, and the household budget just went into the red. You threaten the kids with all kinds of dire consequences. Every week you see the results and yell abuse or praise to the kids and spouse (without giving away any data). What result can you expect? How will morale be in that house?

A Lean approach might be to present the old budget and the new income (if the kids are old enough). Highlight the gap and ask for suggestions to bridge that gap. Everybody participates in brainstorming ideas to run the household on less money without sacrificing their quality of life. Every week you make a chart of the results from the prior week. What result can you expect? How will morale be in that house?

Is There a Right and Wrong Way to Lean?

How you get there makes a difference

Can budget cuts ever lead to Lean Maintenance? This is an important question because many maintenance departments face the budget knife. Lean maintenance is attitudinal so it is difficult to imagine that a company would arrive there from just top-down budget cuts. To achieve a Lean focus, a maintenance department would have to be motivated by the budget cuts. There may be exceptions, perhaps where a leader wants to take on Lean Maintenance in the face of 'no support' from management. That situation is unusual but not unknown.

Cost cutting ≠ Lean

Lean maintenance and cost savings (or cost cutting) are not the same. If Lean maintenance activity was a circle, and cost-cutting was also a circle, there would be some overlap and the two activities would intersect. Within that intersection, Lean does promise savings. Of course, it is necessary to be rigorous about the kind, amount, and duration of any money or time saved.

One of the issues is that maintenance has direct costs and other costs that are the result of maintenance problems. These two types of costs are structured like an iceberg. The direct costs of maintenance are what we see of the iceberg. We call these costs above-the-waterline. Above the waterline, costs are generally smaller than below-the-waterline costs by a factor of 10 or more. Above-the-waterline costs might include inside parts and labor, outside parts and labor, equipment rentals, and several types of overheads.

The Maintenance effort also has huge impacts on the costs and quantities of production for both good and ill. As noted, these costs are referred to as below the waterline, and they include downtime, scrap, process energy, operations labor, etc.

Short term thinking works against Lean

This battle is about short term thinking. This is a paradox because we know that Lean is supposed to be a short term tactic and not a long term strategy. W.E. Deming, the quality expert who is credited with the transformation of Japanese production quality, listed short term thinking as one of the diseases in the pursuit of quality. Like quality, maintenance requires long-term thought horizons. In many plants, maintenance requires that some deterioration be followed for decades and that mitigation be planned years in advance. Most major assets take time to deteriorate, and many give adequate indications of problems. The Lean Maintenance tactic then lives inside a long term strategic approach toward maintenance.

Public companies' stock prices are tightly tied to quarterly performance. Analysts comb over the performance numbers to look for problem areas or opportunities. They look at margins, and determine if trends are good or bad. Top management has bonuses tied to profits and stock prices that are re-calculated every quarter.

Any trend toward increasing maintenance costs is viewed as bad, even if it is reflective of an appropriate, long-term, strategy of asset protection. A few companies make a practice of improving their stock prices by short-term and wholesale reduction of maintenance costs. The fact that these same companies have production, safety, and sometimes environmental problems a few years later, is overlooked or more accurately forgotten.

The best example of this short-sighted mode of operating occurred during the sale of Conrail. Conrail was the government-sponsored successor to many of the old freight railroads in the US (AMTRAK was the passenger part of the business, and as of this writing is still quasi-governmental). After forming Conrail from several bankrupt railroads, the Government spent a good deal of money upgrading the railroad tracks and rolling stock. After more than a decade of independence, management decided that the Railroad Industry was strong enough to sell Conrail.

In the year before they publicly announced the decision to sell, management dramatically cut maintenance personnel and associated expenditures to the bone and beyond. That year, they had their first profit. Reports from people in the field showed that minor derailments, and track problems had increased, and contractor oversight had deteriorated as a result of the cuts.

Is There a Right and Wrong Way to Lean?

The auction for the assets of Conrail was a wild success. With a profitable year, the assets were viewed as much more valuable. After bidding billions, CSX and Norfolk Southern split up Conrail. It is to be hoped that the buyers knew what they were getting, and were ready to restore maintenance spending.

The dark side of Lean

Ricky Smith and Bruce Hawkins in their book *Lean Maintenance,* say that fear often accompanies the lean program. "Many plants undertaking the Lean approach seem to instill fear in their employees almost immediately. Their interpretation of Lean is that productivity must be increased, using the fewest possible employees. To them, Lean means a lean work force, one that will be achieved through fault-finding, blame, and resulting lay-offs." They go on to say that "the fundamental rule of Lean is that a worker who is rendered unnecessary as a result of efficiency gains cannot be laid off."

This is Lean's dark side and it can be seen when companies put Leanness over everything else. The striving for gold overwhelms all other instincts. We may have an example of this with the current crop of problems that Wal-Mart is having. Wal-Mart is, and was by far, the leanest organization in areas of distribution and marketing. Sam Walton, the founder, fought for the lowest costs of doing business. When Sam was alive there were few protests, lawsuits, and negative op-ed pieces about Wal-Mart's labor practices. You got the impression that he cared about the sales associates (whether he did or not). When he visited a store he always knew people's names (and he visited stores 175 days a year until his death). Rumor has it that many of the early employees retired as millionaires, due to the appreciation in their holdings of Wal-Mart stock. This group of millionaires is supposedly larger than the similar group of early retirees at Microsoft.

What accounts for Wal-Mart's alleged abuses against women, overtime, breaks, scheduling, and procurement practices? I submit that the problems are related to the dark side of Lean, with the blind application of Lean without the rest of Sam Walton's vision. All that was left of the vision was the relentless drive for leanness. Sam seemed to have a vision that included the employees welfare, and his vision was the heart of the enterprise. After his passing his successors moved in and made more rules, thinking that it was the rules that made the company great. Without Sam's vision, the company has no moral or ethical rudder, and without a rudder the rule makers (any of us without a vision) run amok. The drive for leanness embodied in the corporate practices, metrics, and structures, are still in place but without the tempering of Sam Walton's vision.

This subject leads to a related conversation on whether an organization can be both huge and Lean? Certainly Wal-Mart is still pretty lean and huge. So the two qualities are not mutually exclusive. But, as with the discussion above about some of the challenges that Wal-Mart faces, it is tough to stay dedicated to lean as an operational philosophy without going to the dark side of lean.

CHAPTER

3

Economics

The most powerful argument in favor of pursuing a project is that it will save real money (unless of course it will also save lives).

In taking on such a challenge, make sure you have a good understanding of the requirements of how your company calculates savings, and at what rate they will green-light a project. Learn the ROI and payback requirements, and how they will be calculated.

Return on Investment (ROI) is the most commonly-used measure for investments. ROI is expressed as the percentage of return earned per year. If the yearly income varies, each year can be evaluated separately, or the years can be averaged together (see ARR below).

ROI of common investments:

Savings account	3%
Money Market	5%
Mutual Fund	12%
Small corporate investment	50%
Capital improvements	30%

Formula: ROI (Return on Investment) = Yearly Income / Total Investment

Example: Replacing 2000 old-style fluorescent fixtures in a school with new technology and electronic ballasts requires an investment of $150 per fixture, or $300,000. The reduction in energy, and costs of ballast replacement and lamp replacement, will yield a savings of $75,000 per year. The ROI calculation is:

ROI = 25% = 75,000 / 300,000   25% is below the school's standard ROI of 33%

Based on these numbers, the school could not justify the investment until it contacted the utility company, which offered a rebate of $45 per fixture or

Rebate 2000 x $45 = $90,000

Recalculation taking the rebate into account results in a new ROI calculation of:

$210,000 (new investment) = $300,000 (old investment) – $90,000 (rebate)

33% = $75,000 / $210,000 The school was able to make this investment because, with the rebate, the ROI met the guideline.

Average Rate of Return (ARR): This calculation is the same as for the Return on Investment (ROI), but extended over the entire life of the investment. The ROI will vary from year to year, but all the returns and all the investments are added together to determine the ARR.

ARR Formula: ARR = Average Yearly Income after Tax / Investment over life

Example:

Springfield Controls purchased a small alternative-energy source (say a hydro-electric generator on a nearby river), for a total investment of $1.4 million. This investment was paid for entirely with internal funds. The average net income (after all expenses and taxes) over the years of the analysis was $210,000:

ARR = $210,000 (average income) / $1,400,000 (total investment)
ARR = 15%

It's interesting to note that, by borrowing, you can sometimes significantly improve the ARR (or ROI). Why is this true? Consider the impact of borrowing funds, where the rate you pay is below your ARR or ROI requirement. The organization will earn a return on the borrowed money equal to the spread between the ARR and the loan interest rate. The US government supports this type of decision by allowing the interest to be deducted from the organization's income tax bill! As companies found out in the late 1980's and early 1990's, and again in the late 200Xs, there is significant risk in excessive debt because payments must continue to be made, even if sales go down and profits evaporate.

The ROI can also be improved if the government or utilities have rebates or tax credits for certain types of investments (such as alternative energy). In addition, in most states, a small turbine can be hooked up to a utility network so that, if your plant is closed in the evenings, and the power output exceeds the closed plant's requirements, you can sell the excess power back to the utility (effectively running the meter backwards). That income also can improve the ROI.

Payback method: The second most common method of evaluating investments is to determine the number of years (or months) it will take to pay off your investment based on the investment's return. The payback method is frequently used along with ROI, and is the reciprocal thereof.

Formula: Payback in years = Total investment / Yearly Income from the investment

Organizations are vitally interested in how soon their money will come back. In the re-lamping example above, the rebate will improve the payback time from four years to three years.

Interesting savings calculation (phantom unless you can monetize it)

Quick calculation of ROI from time savings:

Consider a project to move the parts room. It is clear that repositioning the parts room would save time, but would it save enough time to make the change worth while? We could estimate the savings in minutes per day for the crew. Let's say we could reasonably see 20 minutes savings per day, per mechanic. If we have 17 mechanics they will save 340 minutes or 5.67 hours per day.

Chapter 3

There is a universal formula to calculate Return on Investment in any currency with any labor rate, from labor savings in minutes! There are two simple assumptions:

1. Any savings should generate a 50% ROI or provide a 2-year payback.
2. The total cost of a worker is 2.5 times their hourly wage. This cost includes wages, benefits, overtime, all leave and vacations, lost time, and all the overhead of that worker.

Answer:

Savings in Local currency = Labor rate per hour in local currency x savings in minutes x **20**

So in our example US$20 / hour * 340 minutes a day savings in minutes x 20 = $136,000

Here is the arithmetic behind that quick calculation:
Assuming that 2 years is about 440 days of work

((Labor Rate x 2.5) x 440) / 60 = 18.33 (Cost per minute for 2 years. Round up to 20) x Labor rate

If you do the algebra, pull out the Labor rate and since all that's left is multiplication

(2.5 x 440) / 60 = 18.33. Put the labor rate back in and that's the savings per minute. Pretty easy.

Economic Modeling: How do you know an alternative is Lean?

Economic modeling will help you determine the Leanest alternative (economically speaking). Economic modeling simulates the costs and income from a particular maintenance alternative, given the economic 'facts' of the case. Models can be as simple as projecting the costs and income, to sophisticated models that include interest rates, tax policies, and other variables. If the consequences warrant it, alternatives can be analyzed using economic modeling.

Economics

Many different strategies can be used in maintaining particular assets. Of course, the first choice is to employ an asset that needs no maintenance! If no maintenance is possible, look at that alternative before 'settling' on a PM or another maintenance alternative.

Each choice has both economic and non-economic consequences. Economics is important, but other issues might suggest a particular strategy. For example, PM might be the best strategy for a particular asset but your company has no PM system and a bad track record of allowing downtime for PM, so PM will not work. There are 6 or 7 major strategies, and many more combinations and sub-strategies. A few of them could be:

- Run the unit in breakdown mode–where it gets no attention unless broken. This example is always used for comparisons. In modeling, doing nothing should always be a choice (and sometimes it's the best choice).
- Redesign the asset to be quick-connected in place, so that you can do a quick switch upon failure and rebuild it off-line for use when the next one fails. Although the failure rate will be the same, each failure would be handled more quickly.
- Design a basic PM program with inspections, basic maintenance, and occasional as-required corrective jobs.
- Just bite the bullet and install a whole Back-up (say a pump) that can be switched over quickly before any chargeable downtime is incurred.
- Planning for component replacement with quick connections is like the second choice above, except that the asset is swapped on a scheduled basis before failure.

Example: Repeated failure of a chemical transfer pump has been occurring for the last several years. Downtime from lost production is valued by the cost accounting department at $500 per hour after the first hour (no cost for the first hour). There is a reservoir that will run for 1 hour before the downstream process shuts down. Labor hours are valued at $40.

Engineering analysis shows that the application is severe and that this performance is the best that can be expected (eliminating the no-maintenance alternative). The skilled mechanics working with engineering have designed a PM task list that will drop the number of failures dramatically.

Economic model for breakdown mode

Currently, in **breakdown mode,** the pump is failing 4 times a year. Each incident requires 10 labor hours and $2000 of parts to get the pump back on line. Downtime from calling maintenance to full operation is 14 hours (2 hours to respond to the call and 2 hours to get the system filled up and back in operation. A mechanic is required for only 10 of those 14 hours. It is a 1-person job).

In this model, the only thing required from management is to keep the spare parts in stock. A simple min-max system would suffice, with inventory levels depending on the lead time.

Economics of breakdown maintenance

Annual Breakdown costs =	Probability of breakdown in any year (how many times it fails)	(Cost of breakdown =	Cost of Downtime)
		(Labor cost + Parts)	(Downtime-1 hour) x downtime cost per hour
$35,600 = 4 x $8900	4	$2400 = ($40 x 10) + $2000	$6500 = 13 hours x $500
	4	$2400 +	$6500

The economics of this choice are clear, but what are the other consequences? There are several kinds of consequences. For one, the customer (operations) is disturbed 4 times a year for over one shift each time, amounting to 52 hours of chargeable downtime a year. This problem is not too bad if you have only one such pump, but imagine if you had 100 pumps, each failing for over one shift, 4 times a year. Of course the pumps would break at the least convenient times, and probably the failures would bunch up (lengthening the time to respond and the downtime).

Economics

Running in breakdown mode requires 40 hours of maintenance worker time. Another consequence would be the cost of $35,600 per year (in both above- and below-the-waterline costs). On the plus side, this situation requires little management and no capital expenditures. Although safety and environmental considerations are beyond the scope of this model, it is intuitively obvious that unscheduled and random events such as breakdowns are the least safe choice.

Economic model for hot swapping

Another breakdown alternative is to **replace the pump** after failure with a rebuilt pump and rebuild the faulty unit off-line. In the computer field, this approach might be called hot-swapping. Components (such as circuit boards) are swapped with the computer running. We can't actually swap the pump while it is running but we still call the method hot-swapping (maybe we could call it warm swapping). To set up the method we would have to purchase a second full-pump unit and adapt it for quick changes.

The second pump will cost $15,000. After engineering the quick change, swapping to the second pump will take 2 hours. Downtime is down to 6 hours (2 hours to respond, 2 hours to swap and 2 hours to fill the system). The rebuild will cost the same amount as the breakdown model above, $2000 for parts and 10 hours off-line. Only 1 pump core is needed to rotate with the installed pump so we have to be sure to charge for only 1 pump core. The purchased pump is a capital expenditure.

Management needs to ensure that when the damaged pump is removed from service it is rebuilt reliably within 2–4 weeks. That element of management is in addition to a simple stocking strategy like the one described above. If the pump is not rebuilt in time, you run the risk of spending money for the spare and still having the downtime of the breakdown option. Having a breakdown on top of not having the pump rebuilt is the worst of both worlds (and is fat maintenance not Lean Maintenance).

Chapter 3

Economics of hot swapping

Breakdown costs =	Probability of breakdown in any year x	(Cost of breakdown + (Labor x Labor rate) + Materials	Cost of Downtime) (Downtime-1) x $500
(Plus 1-time $15,000 capital investment for spare pump)	4 (same as breakdown)	Labor: 2 hrs to change pump + 8 hours to rebuild = 10 hours x $40.00/hr = $400 labor + Materials = $2000	6 hours (1st one free) = $2500 = 5hr x $500/hr
$19,600 = 4 x $4900	4	Total Incident Cost = $4900 = $2000 + $400 + $2500	

What are the consequences of this choice? Once again there are several kinds of consequences. The Operations or production process is disturbed 4 times a year but this time it is for less than one shift. Chargeable downtime drops by 24 hours per year, or better than a 50% reduction, which is an improvement. There is no difference in the cost of maintenance labor or parts for this scenario. Annual costs have improved to $19,600 because of the reduction in downtime. The costs above the water line are the same. Each incident will still need 40 hours of maintenance labor and $2000 of parts. There is a capital cost, but it gets paid off within the year. More management is needed than with the previous example, to insure that the pump is rebuilt in a timely way. Safety is not affected because the failure mode is still random and unscheduled.

PM

PM is a common strategy. In fact, it might be tough to run a breakdown-only scenario because most firms do some PM (even if it is only a shot of grease). Frequently we notice that the greatest returns come from the most basic PM activity. There is a diminishing return to increased PM expenditure. Of course, if the subject is a fuel pump on a jet engine, the added benefit is worth while.

Economics

The effectiveness of the PM is always at issue and is discussed in another section. An ineffective PM will consume resources without providing an adequate Return on Investment. It is important to look deeply at different strategies because the best way to take care of an asset might depend on the downtime cost, not just the breakdown cost.

The **PM routine** was designed by the mechanics, will take 1 hour a week, and requires downtime to accomplish (but that hour of downtime is free). We will assume that the PM routine as designed is effective. Additional corrective repairs (resulting from the PM inspector finding deterioration and fixing it before failure) take 5 hours per incident (3 incidents per year on average) and $1700 worth of materials per incident. With the new PM program, the system becomes significantly more reliable, and breakdowns will drop to 1 every other year (each breakdown costing the same as the original breakdown mode. that is: 10 labor hours + $2000 worth of parts and 14 hours of downtime).

Economics of PM

Total Annual Cost for PM case	(Cost per service x	# Of services per year) +	(Cost of corrective repairs with downtime) +	(New probability of break-down x	(Cost of break-down +	Cost of Downtime)
$21,230	$40	52 services per year	Corrective Work ($1700 mat + $200 labor + $3000 DT) = $4900	.5	$2000 + $400	13 x $500
	$2080 = $40 x 52		$14,700 = $4900 x 3 Incidents	$4450 = 0.5 x ($2400 + $6500)		
	$2080 +		$14,700 +	$4450		

What are the consequences of a PM strategy? The customer is disturbed 3½ times a year. The significant difference is that 3 of the interruptions are scheduled. Operations can choose when to go down for the corrective work. Overall, the chargeable downtime is about the same as with the hot-swap example, and provides more than a 50% improvement over the breakdown mode.

One of the biggest changes is that the model calls for 72 hours of labor, or a 55% increase. The above-the-waterline costs are higher, and the total annual costs are slightly higher than hot swap but still well below Breakdown. Again, with one pump the impact is not noticeable. A hundred pumps would require 3200 extra hours, or about 2 people full-time. There is no capital cost.

Much more management (perhaps a CMMS PM module) is needed than with either of the two last examples to ensure that the PM is done weekly and the corrective maintenance is done in a timely way. The stocking system has to ensure that spares are on the shelf when needed. Safety is improved because of reduced non-scheduled events.

PCR

Another PM alternative is Planned Component Replacement, which has two versions: Planned rebuild (as in the previous example), and planned discard (used for low-cost components). PCR is closely related to the quick-change alternative except that the pump is changed out on a planned basis before failure.

PCR was one of the primary strategies of the aircraft field as well as in Air Forces around the world. Advantages include the ability to schedule technology upgrades to the equipment. Good scheduling practices are encouraged by allowing accurate workloads to be determined for an entire year, and longer for major overhauls. PCR is expensive, and is usually only justified where the consequences of the breakdown are expensive, dangerous, or both. Newer approaches toward higher levels of intrinsic reliability and advanced Predictive Maintenance inspections have forced PCR into a back-seat role in its traditional industries.

In our ongoing example, a PCR interval of 2 months would be required. The pump would be changed every two months, whether it needs it or not. The PCR operation (as with the hot-swapping operation) would take 2 hours of mechanics' time. Bringing the pump back to operational specifications would take 5 hours on the bench each time, plus $500 worth of materials. One other advantage is that PCR can result in extremely high levels of reliability. The new failure rate would be once in 10 years (with costs similar to those of the breakdown example).

PCR Cost =	(Cost per PCR x	# Of replacements per year) +	(New probability of breakdown x	(Cost of breakdown +	Cost of downtime))
(Plus a one time capital cost of $15,000 for spare pump)	$2280 = (3hr x $500 Downtime) + (2hr swap + 5hr repair x $40) + $500 Mat	6 per Year	Once in 10 years or 0.1	$2400 = $2000 Mat + (10 x $40 Labor)	+ $6500 downtime
$14,570	$13,680 = $2280 x 6		$890 = $8900 x .1		
	$14,570 = $13,680 + $890				

What are the consequences of this choice? Once again, there are several kinds of consequences. Production is disturbed 6.1 times a year for less than half a shift. The interruptions are almost all scheduled. Chargeable downtime drops by 19.3 hours per year, or better than a 20% improvement over hot swapping, which is a large reduction. There is also a 3-hour increase in maintenance labor with this mode. Annual costs are improved to $14,570 because of the reduction in both above-the-waterline costs and downtime. There is a $15,000 capital cost, but it is paid off in less than a year.

As mentioned previously, more management is needed to ensure that the PCR is done on schedule before failure, and that the pump is rebuilt in a timely way. Safety is improved because there are few unscheduled events.

Back-up

Many organizations choose to operate critical processes with one or more back-ups. Back-ups are all around us in critical systems.

Chapter 3

Few people would be inclined to fly globally if jets still had one engine and no way to fly when it broke. Most hospitals of any size have back-up power generation to replace electricity in a utility outage. The downside to building plants with back-up pumps and compressors come in operational and economic areas.

In the operational area it is thought that having back-ups makes the maintenance department sloppy and removes the urgency from the equation. If we have one compressor we are extremely focused in keeping it going. Having two or more compressors take away the excitement from a breakdown. One of the upcoming key performance indicators of manufacturers is Return on Assets. Return on assets goes down as the asset base goes up (for the same profit level). Of course the reason for the back-up is to enhance and stabilize the production and the profit levels too.

An alternative strategy is to mount a **back-up pump** in the system with an automatic alarm. We assume there is enough room to fit a back-up in place. The operators can then switch the pumping load to the back up when needed. The cost of the back up and associated piping and controls is $25,000, including labor, and will take 20 hours of work time (one time fee). A back up can be switched on without downtime. Failure rates may be assumed to be the same as in a breakdown, but without downtime. The cost of the back-up pump is charged only once, and it should be capitalized, depreciated, and held in an asset account.

Breakdown costs =	Number of breakdowns a year	(Cost of breakdown +	Cost of Downtime)
One time cost of back-up $25,000 + (19 x $500 = $9500) = $34,500	4 incidents	Annual Repair costs $2400 = $2000 Mat + (10 hours x $40)	$0
$9600	$9600 = $2400 x 4	$2400 per incident	

Economics

What are the consequences of a back-up strategy? Customers are happy because they are never disturbed (which is why this set up is so popular). After installation, chargeable downtime is zero. Labor and parts (above the waterline), are the same as for the breakdown model, and annual costs are the lowest, though there is high capital cost. Not much management is needed beyond being sure the unit is repaired after failure and put back in service, to be ready for the next breakdown. Safety considerations are the same as for the breakdown model, (not great).

Here is what you should take away from modeling: Modeling is a Lean activity. It is one of the few ways you can sit in an office with a spreadsheet and have an impact on maintenance and operating costs for the next twenty years.

Every maintenance alternative that can be imagined costs money, and every alternative has consequences. Pick the alternative that gives the least costs and the consequences you want!

CHAPTER

4

Lean Maintenance and World Class Maintenance

World Class Maintenance is an extremely useful fiction, invented to spur companies toward excellence in their respective fields. It is important to realize that there is no one World Class standard. What is World Class in a nuclear power plant would bankrupt a chicken-processing plant. World Class is always referential toward one or a series of related industries. For example, oil refineries and large chemical plants might be within one World Class division. Chicken processors and other meat processors might be in another. The rules and benchmarks could be compared within the division.

Having said that, if there was a world class standard it would relate to its attitude toward customers. Womack and Jones outline a World Class approach to supplying customers in *Lean Solutions* that is also Lean. Customers want:
- My problem completely solved
- Don't waste my time
- Provide exactly what I want
- Deliver value where I want it
- Supply value when I want
- Reduce the number of decisions I have to make to solve my problem

There are World Class attitudes that are common across industries. In *Managing Factory Maintenance Second Edition* I reviewed some of these areas. I want to revisit those areas while adding in discussion about their relationship to Lean Maintenance.

Also, since that work was published, the global urgency for efficiency and sustainable production has heightened.

Six Areas of focus for Lean Maintenance in companies that strive for World Class

Lean is both a top down and bottom up activity. It is top down in the sense that the best implementations are driven by a company's ethics and culture. Even high-level personnel are involved in the Lean efforts. These efforts are on manager's lips whenever they speak. Lean can also be thought of as bottom up. In most projects, the personnel closest to the action know intimately where the waste is. After all the workers are the ones to sweep it up, or it is generally their time that is wasted.

Along the way management has developed new attitudes toward maintenance activity. The best organizations realize that, although some maintenance is inevitable, a proper attitude is necessary to minimize their exposure (in other words their downtime, parts costs, and labor costs). They also realize that maintenance has above-the-waterline costs and below-the-waterline costs (that become evident when maintenance is cut too far). The below-the-waterline costs can dwarf the other costs. In short, these organizations view maintenance activity in a particular way. This new view happens also to support Lean initiatives.

- Management wants a maintenance department that is more proactive and less reactive. Reactive maintenance is the fattest of all types of maintenance activity. Finding a problem, planning the solution, and solving it (hopefully once and for all, if possible) , is less expensive in the end than having the problem find you (over and over again). This truism is fundamental to Lean maintenance. Any repetitive unplanned, unscheduled, disruptive, event is by definition Fat.

- Part of the proactive approach is don't wait for breakdowns to mess up production, get out there and find deterioration before failure. The big money is below the water line, so use all techniques and technology to preserve production. Use non-interruptive inspection techniques, or well-timed, scheduled outages (but work diligently to reduce their duration) to detect performance degradation and potential failure points.

- Once something is broken, apply root cause analysis (RCA) to manage the cause, not just the symptom (the failure). Structured RCA is one of the most powerful Lean techniques and will be covered later, in its own section.
- Less repetitive maintenance effort is better (this includes repetitive PM activity)—solve problems permanently where they don't break or even need PM either.

In the old days, managements were cowardly because they hid behind their ignorance.

- They wanted the results of advanced maintenance management practices without the investment. So Management didn't put their money (or authorized downtime) where their mouth was. If they read in a trade magazine that PM was a good idea or PdM was the way to go they asked for PM/PdM without providing adequate financing, support, or in-depth training. They thought that by yelling loud enough, or setting big enough goals, the machines would be motivated to not break.
- What happened was that marketplaces became more global, with tough new competitors, and that left no room for amateurs. Judicious long-term maintenance investment made winners. All of a sudden you find you are in competition with a factory with half your labor rates. It takes courage to face that, do the tough work necessary, and some luck to end up the winner.
- Every part of the company has expertise. Utilizing these islands of expertise is essential for success. Management is listening to maintenance opinions, and factoring those opinions into large business decisions. Using your experts to their best ability is Lean.
- Like any change, if management wants improved maintenance they must fund it, and care about it for a significant period of time.

Maintenance has to face some responsibility also. There is a culture of deception in maintenance because it didn't seem that management could take the truth. In those old days, maintenance folks hid what was really going on in the maintenance department from management.

Lean Maintenance and World Class Maintenance

Now the spotlight is uncomfortable because it is on maintenance, and it is looking for hard numbers. All through the company, hard numbers are king.

- Why? To see what is really going on. Secondly, numbers are used to measure continuous improvement (or lack thereof) and Lean project progress. Remember, hard numbers measure real savings and real improvement, and not phantom savings.
- Productivity is now secondary to results. The focus is simultaneously on doing a thing right and on doing the right thing. In fact, world-class organizations realize that high productivity is a function of doing many small things right. Management really wants results on the larger cost area below the water line (such as uptime, and reliability). The world-class organization's management wants high productivity, low waste, and more sophisticated management of maintenance from the maintenance leadership.
- World-class organizations have an increasing willingness to use sophisticated tools of statistics, finance, and accounting, in maintenance analysis. Sometimes the only way to see the fat is through sophisticated tools and processes.
- Management is starting to require analysis-driven maintenance decision-making, not seat-of-the-pants-driven maintenance decision-making. The analysis can uncover non-obvious Fat. The key is making rational decisions backed by data that can be reviewed (and understood) by a manager without a lifetime of maintenance experience.

Great maintenance managers have realized that how people are used is the key to success. If W.E. Deming taught us anything it was that the system you operate in is sometimes more important than the people, and that the right people are your major asset. Good people with a bad system will be thwarted and demoralized.

- Your system should encourage ad hoc teams to solve problems. Teams solve problems, and are dissolved when the problem is solved. Problems are like the fat you trim off your steak before you cook it.
- Barriers limit thinking. Limited thinking gets in the way of Lean thinking. You can observe the fading of traditional departmental

barriers (letting people see more of the big picture to make better decisions).

- Who knows better what the customer really needs than the customer? Where appropriate, World Class companies encourage customer participation in Maintenance (with training, and with proper management called TPM).
- Information in the right hands makes Lean decisions possible. Information sharing includes things like sharing charge-back rates, machine part costs, and the maintenance budget (for starters).

The world has changed. It moves more quickly now, so that only nimble competitors survive. Nimbleness is a function of the right systems driving the right people. The right people have the right attitude as well as the right competencies. In fact, almost all problems are really people problems masquerading as maintenance problems.

- If people are your only asset then it is prudent to invest in your people through continual training to improve the effectiveness of this asset. Although competence of your people is tough to measure directly, you know when it is not present! When it is present, things run smoothly. Frequently smooth running is Lean.
- Layoffs are the bane of the maintenance department. If you've been following the advice to train people every year, you have a substantial investment in them. It takes years to develop the expertise of a laid-off maintenance worker. We also want to be damn sure that nothing within our control gets in the way of our people's concentration. Fear of layoffs is handled perhaps with a smaller core crew, and supplemented with contract help. Commitment is to people, so that every other option is looked at before layoffs (W.E. Deming says "drive out fear").
- Cross training (also known as multi-skilling) is a goal to improve both productivity and the personal sense of satisfaction. Cross training is under utilized. Due to fear in Union environments and laziness elsewhere it is rare to see concrete programs and incentives for cross training. With proper safeguards (on quality and competence) cross training can boost productivity dramatically. Being able to send one person instead of two is Lean.

- The Attachment is to the people rather than technology or computer systems. CMMS and other systems have come a long way in the last 2 decades but none of the systems can operate without human intelligence driving them.

What is the real mission of your department? In the best organizations there is alignment between the mission over the door and what really is said, or not said, in the small meetings and on the shop floor. Always:
- Focus on service to the customer or focus on adding value to the customer.
- Focus on safe operations for the maintenance workers, operations, the general public, and the environment.
- Look for ways to reduce costs (ongoing Lean Maintenance) of your operation.
- Look for better ways to do business by cultivating willingness to run controlled experiments (can be directed by Lean projects).

Once again, the goal is the same. We are seeking a powerfully self-motivated workforce, and excellent execution of maintenance activity.

Returning to the discussion about specific industrial groupings there are certain business process areas considered in World Class studies that apply within each division of industry. Within these areas we are able to relate the practices between cereal manufacturers to each other, and even cereal plants to similar plants such as snack makers (if they are similar in size, scale, and approach).

Below are some of the strategies of world class maintenance management and how they specifically relate to Lean Maintenance.

☐ Effective store room and stocking system	Essential for a couple of reasons. One is reduced downtime. Another is higher worker productivity, third is having the lowest inventory level to support the assets.

❑ Effective PM and PdM program	Although PM/PdM is not the goal, it is a fundamental part of world class maintenance. The intensity of the program will be related to the consequences of unscheduled failure.
❑ An up to date Maintenance Technical Library (elements can be either virtual or real)	World Class organizations manage information effectively. Lean Maintenance requires that 1000s of details are on hand when needed.
❑ Timely reporting of potential problems by production, and meaningful feedback to planners on completed jobs by supervisors and technicians. Open dialogue with operators and engineering on troublesome equipment	Communication is the core of World Class Maintenance. All parties communicate to their critical partnership person. All communications channels are open.
❑ Outsourcing and proper use of and orientation toward contractors	Use of outsiders is indicated when there are not enough available hours of competent personnel or there are no personnel with the competence needed at all. It is also used to minimize overall cost to deliver the maintenance product. Good for seasonal or business cycle peaks.
❑ Good use of a CMMS in an organization running multiple shifts or locations, or who have more than 20 workers	The computer is a great tool for all but the smallest maintenance organizations. It provides the data for Lean projects and any World Class department would be using one intensively (not just feeding it but also taking advantage of it).

Lean Maintenance and World Class Maintenance

❑ Complete equipment repair history	The CMMS or a good manual filing system provide the data for Lean decisions and are essential in a World Class setting. Data is retained from date-in-service until retirement.
❑ Training	So much of the continued success of your World Class maintenance effort depends on things that you do not know today. Training has to be part of the World Class program.
❑ Thorough failure analysis, Root Failure Analysis	World Class organizations do not tolerate repetitive failures. They have exercised the muscle (possibly through Lean projects) to analyze and redesign or permanently repair, all repetitive problems.
❑ Disciplined approach to production such as understanding the importance of sticking to a schedule in both production and in maintenance	World Class maintenance is not only a maintenance 'thing.' It requires discipline and cooperation from other groups. That cooperation facilitates Lean projects.
❑ Willingness	A World Class maintenance function has the willingness to embark on programs that make sense such as RCM and TPM, to enhance the delivery of service and reduce cost.
❑ Quality	World Class maintenance would be nowhere without a sound and coherent approach to quality. Quality of maintenance work is related to proper training (competence), proper tools and parts (bill of materials) and enough time (custody). All these same elements point to a Lean organization too.

❑ Good workmanship by craft personnel. Existence of overhaul and rebuild capabilities	The core of the whole discussion is excellence, without which World Class maintenance cannot exist. Lean projects sometimes require deep capabilities that are either in-house or easily available locally.
❑ Good use of repair technology	World Class Maintenance is not about all the new toys introduced by vendors. It is about knowing what is available and using the appropriate technology when that is the best alternative.
❑ Good relationships with vendors	If selected intelligently your vendors can be used to solve some of your most intractable problems. All World Class organizations have partnerships with their vendors. This multiplies their expertise without significant extra cost.

Lean Workers

*Benjamin Franklin says: An investment in knowledge pays
the best interest.*

When you think about lean workers and lean workforces, think about the difference between a platoon of army or even marine personnel, and an equal number of Special Forces personnel. Special Forces have the ability to focus high levels of firepower for short to moderate periods of time, on an objective. They can move quickly because they are designed to be maximally mobile. The Special Forces have deep ability to be flexible and adapt to the situation. Finally, they move without disrupting the environment (stealthy).

The focus, speed, stealth, and flexibility bring about the objective. The lean maintenance worker is like that Special Forces soldier, flexible, with the ability to focus adequate firepower on any maintenance problem.

It is very simple to define an ideal maintenance worker from a Lean perspective. The ideal workers are those who have the skills to handle whatever maintenance tasks are thrown at them. Taking that concept to its extreme, the most efficient set-up is where all roles are in one person. That arrangement gives maximum flexibility. When we speak about all roles, we mean all: maintenance, production, set-up, QC, and so on.

Did you ever notice that, early in the history of almost every company, there is a stage where everyone does everything? Companies in that stage are most Lean because there is very little surplus anything. Everyone moves toward the problem. Problems get handled (or the company goes under!). With too much fat, the company runs out of money.

Yet specialization has many advantages, both in the factory and even in society as a whole. When you specialize you become an expert at something, and you can work more efficiently, Society then has surpluses over having one person do everything. Surplus comes from each person focusing on one thing and gaining the mastery and efficiency that cannot be reached by a generalist.

In our world, if you feel sick, you might see a generalist (internist) who has a wide but shallow knowledge base. If you need a coronary bypass you had better go to a sub-specialist, and not even a general surgeon. The same thing applies to maintenance. I would hope our welders are pretty well cross-trained with all kinds of welding. Code welding in a Nuclear power plant is a specialty, and if we have that kind of work we want a specialist to do it.

Perhaps there is a middle ground. Lean workers are workers whose skills are used well. It would be fat to have workers with skills that are unused by custom, contract, or ignorance. For example if we had that code welder working in his or her specialty once a year that would be fat. On the other hand, when an operator does TPM maintenance, the company moves toward Lean. Certainly, when an operator jumps in and lends a hand to the maintenance worker during a repair, it is Lean. Even when a maintenance worker jumps on the line to keep it running during a break it is Lean.

Some things to look for to Lean up a company's work force

Continuous training is essential. Training time per year, on average, ranges from 1% (about 20 hours) to 5% (about 100 hours) of technicians' direct hours per year, every year. If the goal is multi-skilling or cross-training, the amount of craft training might be quite a bit higher particularly during the transition. The goal is not to make everybody know everything, but to maximize each worker's effectiveness when they are on jobs. An example would be to teach the millwrights to safely wire and unwire motors, where there are no known problems. This training increases the effectiveness and Leaness of the millwrights.

Training money is budgeted and there is a KPI to ensure it is being used. The best organizations make a determination of benefit, or return on investment (as best they can), from all training activity.

Look deeply at the goals and the controlling culture of your company. Is cross-training a goal for the department, to allow craftspeople to do the `whole job and provide scheduling flexibility? The more often the maintenance worker can do the whole job, the leaner the department is operating. There was an excellent story about this issue at a major metropolitan city airport.

There are rules about the crafts that have to be in house when an airport is operating. Some of these rules are derived from the FAA, and others develop locally. Whenever this airport was open, among the required personnel there needed to be a licensed electrician, a licensed elevator/escalator technician, and an expert on baggage equipment, among others. The total was about 23 people. The airport was deemed to be open 24 hours a day, because freight operations ran at night. These jobs were traditionally filled by different individuals, but a new contractor got the FMC (Facilities Maintenance Contract) and his manning schedule had one person covering all three roles. This individual had dual licenses in Electrical and Elevator/Escalator, and had demonstrated skills on the baggage equipment.

When the smoke settled, and the crewing plan was approved, the manning requirements in the middle of the night had been reduced from 23 people to 7 or 8. The company certainly could crew up if there was specific work to be done (most airport work was done at night), but they weren't compelled to (such as on the weekends).

Not everyone is created equal in maintenance skills. In many parts of the world it is very difficult to get people with solid maintenance skills in any craft, let alone in more than one craft. Does your company have a consistent process to attract, identify, and hire, the best maintenance workers? Are the wages competitive? Does the maintenance department have either the final decision or major input into the hiring decisions?

Chapter 5

Lean workforces can be explicit goals. To achieve the goal, the technology and skill sets needed to support the equipment have been analyzed and discussed. There is an ongoing assessment of the skill sets of the existing workforce. The skills of the workforce are compared with the skills needed to maintain the equipment at that stage in its life cycle (new plan or old plant). The changes budgeted for the next few years are included in the assessment. Gaps in skills are filled by hiring and training.

Lean idea for action #1: The best time to negotiate training from vendors for new equipment is BEFORE signing the contract. Be sure that training for the maintenance department is part of all new equipment acquisition contracts. To be even more clever, negotiate training days with no expiration dates. In a few years, when the asset starts to wear out, you can send people away for training that can be used immediately.

Lean idea for action #2: Ask all machinery vendors if they have any free (or low cost) O&M videos. Many machine builders produce videos about how to set up or fix their machines, and those videos can provide instant training. (The videos should be examined prior to showing, to ensure that they are useful).

How to Lean up Training

The role of training in Lean is essential. Only people who are up-to-date in their craft can make the maximum contribution to Lean efforts. There is another, more subtle effect of training. It makes people more flexible in their attitudes and approaches to problems. Both attributes are important in Lean maintenance.

The skills needed to run today's factories and buildings are changing faster than people can assimilate. If your average maintenance workers are 40, then it has been 20 years since they got the bulk of their formal training. Think of what they were learning about 20 years ago and look at your equipment. These jumps in technology disorient even the most dedicated worker. You are in the training business, and you might as well be good at it.

Lean Workers

Attitude, Aptitude, Ignorance

Before embarking on an elaborate training program, ask yourself the question, what is the cause for lack of performance? If you think you have an answer, how do you know? Is there evidence, or do you just have suspicions? You might need some support to determine whether you are facing an attitude, aptitude, or ignorance issue. The strategies for dealing with each are quite different.

There are three general causes for inadequate performance:
Attitude problems
Aptitude deficiencies
Ignorance, or lack of knowledge

The core of the question is, does this person need training, or some kind of counseling, or are they unsuited or unable to do the job? Some performance problems come from a bad personal attitude. However, bad attitudes are thought to be more common then they actually are. Studies show that 92% of performance problems come from either inadequate training or lack of ability such as strength, reach, or intelligence (which cannot be overcome even with practice and training, unless you are willing and able to accommodate the job to the capabilities of the individual). To make the diagnosis decision more difficult, many people develop attitude problems as a defense mechanism against feelings of ignorance or incompetence.

Competencies

In training terms there are three types of learning (called competencies) that apply to maintenance; Knowledge, Skill, and Attitude. For higher-level jobs (such as chief service person), to be effective, a person must be competent in all three. Many types of training address one or other of these types of learning without regard for the others. Maximum effectiveness must cover all three areas. There has been a shift in the way these competencies have been looked at in maintenance that should be examined carefully before training is designed.

Chapter 5

Traditionally, the maintenance field was oriented toward evaluation of specific skill sets. Skills are still important but, particularly when doing service calls, knowledge also becomes important.

TYPE	NOTE	OBSERVABLE BEHAVIOR	PERFORMANCE LEVEL
Knowledge Learn in class, books	Usually was the area of engineers	be able to describe, diagram, argue, etc.	answer X of 10 questions correctly
Skills Learn by doing	The core of maintenance	demonstrate, show, perform, solve	do... in x minutes with no mistakes
Attitude Learn by mastery	Almost ignored by maintenance	comfort, without hesitation	To your own satisfaction

There are five steps in the design and delivery of a tailored Lean learning program:

Pre-step: be sure you are looking at a training issue (and not an aptitude or attitude issue), as discussed above.

Step 1: Determine what knowledge, skills, and attitudes are needed for the job. Before we can look into teaching anything, we have to see what is needed. Look at the job as it is today, and forecast where the job is going in the short term. The big picture of competencies is called the General Learning Objective (GLO). The concrete and specific skills, knowledge, and attitudes required to do the job, are called specific learning objectives or SLOs. If the job is properly deconstructed, a person achieving these SLOs would be successful in the specific job (in training terms he or she would have the competencies to do the job).

If the training is for a bench service technician, we have to look at what competencies are needed to be successful. What (knowledge, skills, and attitudes) is needed to trouble-shoot microprocessor boards?

Lean Workers

It is necessary to decide what level of competence is appropriate for a service technician in servicing these boards.

Step 2: Evaluate the potential trainee's (or trainee groups') current skills, knowledge, and attitudes. A direct supervisor might be able to make an educated guess. If the trainee has good insight, he/she might know where they are weak. Most situations require some kind of testing (either observation on the job or more formal written or bench tests). The testing should be designed to uncover the skills on the "required" list.

It is important to note that success on the test should correspond to success on the job. Testing that does not reflect job requirements is said to be invalid. In the US, legislation is clear that the test must not discriminate against any group, disability, or condition. For example, if the worker must lift 100 lb. in the test, the job must call for heavy lifts where equipment cannot easily be used.

We must evaluate any service-tech candidate's skills, knowledge, and attitudes. After we make a list of specifics we rate him/her (or test) in each area. The result would be tailored to the specific individual. The testing uncovers the SLOs (specific learning objectives), where there is an issue.

Step 3: Develop the training program to fill the specific gaps. To design the training, translate the voids in skills, knowledge, and attitudes of the potential trainee from the "required" list to develop a training lesson plan. The training plan should list all the types of learning that this person/group needs. The plan summarizes the skills, attitudes, and knowledge that the specific candidate lacks, but needs for the job. At this point we would estimate the time requirement for the trainee and the requirement for any supporting staff. The training would be designed to be appropriate to the competency being trained. For example, no amount of classroom time will effectively train people in skills (unless the skill being taught is teaching).

Step 4: Deliver the training. Use inside and outside resources for trainers. Work to reinforce any training with relevant work, during or immediately after, any training. Putting training to work immediately is essential.

Step 5: Assess whether the candidate learned all the skills and knowledge, and adopted the attitudes needed from the training, to be successful on the job? In simplified language, was the training successful? If it wasn't, then the candidate must be retrained. Retraining usually means that you have to go back to at least step 2 in the process, and determine the voids in skills, knowledge, or attitudes.

You might go through this same exercise for all related jobs. A senior service technician might have related SLOs (and others beyond this set) that can be incorporated economically into this training. A field technician might only need to know a subset of these items.

Lean maintenance makes sure the right people are doing the right jobs

Do any of the maintenance people enjoy mixing it up with the worst breakdowns, can improvise, hate rules and paperwork, and enjoy the attention that comes from fixing high-visibility breakdowns under the gun? What about anyone who is methodical, enjoys structure, and hates variations in their routine?

A Lean Maintenance manager looks at the people and puts them in the places where they can grow, prosper, and be happy, and where they can make a maximum contribution to the success of the company. Of course, Lean means flexible, so everyone might find themselves doing all kinds of tasks.

There are three basic types of work in a maintenance department. In addition to work assignment by craft or skill set, attempts are made to allocate assignments by work type.

The **PPM (PM/PdM together called PPM) group** provides reliable PM, PdM and Condition-based maintenance services. This person/group concentrates on preventive/predictive maintenance work. The group consists of people who can work without close supervision, like the lists and paperwork, and can tolerate the boredom of PM activity without losing their observation skills. An optimized PM/PdM process requires 15% to 20% of maintenance labor. Neither Emergency nor Scheduled Backlog work should be allowed to interrupt the work of this Group.

Lean Workers

The **Planned Backlog Relief Group** provides timely relief for those work requests that have adequate lead time for planning. Backlog consists of all plannable-work (non emergency or urgent). This group includes the best mechanics, the most thorough workers, and people who like some variety.

In a pro-active environment, the bulk of the maintenance workload should be plannable. Sixty-five to seventy-five percent (65% to 75%) of maintenance resources should work in this mode. The Planned Group is called upon to support the Emergency Response Group whenever it encounters peak demands (10% of the time by design). This strategy still allows the Planned Group (constituting 65–75% of the maintenance workforce) to be scheduled and assigned to well-prepared jobs and not be interrupted 90% of the time. This organizational mode is far more pro-active than most maintenance departments anywhere.

The **Emergency Group** provides prompt response to true-urgent needs. The group has responsibility for handling essentially all urgent demands; requesting assistance only when necessary. In other words, this group protects the other two groups from interruption. The E group likes variety and attention, is multi-skilled, and inventive. It cannot fulfill its objective for 100% of the time unless it is staffed for peak demand, which is not the optimization sought (not a Lean crewing strategy). The Planned Backlog Relief Crew provides assistance. In a pro-active maintenance environment, prompt response to urgent work calls requires approximately 10% of maintenance labor resources (assuming multi-skilled personnel staff the Emergency Group). Urgent work, by its nature, offers little opportunity to plan or schedule except in a most rudimentary way.

CHAPTER

6

Lean Maintenance and Safety

Lean is making a process run without any activity that is not necessary to support the process. In manufacturing, activities are considered value-added and not value-added. The issue here is, where does safety fit into Lean maintenance? Environmental compliance, or safety of the environment is related to safety. Lean maintenance is completely consistent with high safety, and a responsible attitude toward the environment.

What happens when a company cuts too much money out of maintenance? Immediately, nothing happens. It takes a while to liquidate the good conditions of well-maintained assets. The first signs of the problems coming are subtle. The manufacturing process starts to have increased variation as the assets suffer deterioration. There is also an increase in the workers stress level. The increased stress is the result of having too many jobs for too few workers, and the lack of relief workers. The lack of relief workers means that any small hiccup in attendance crashes the already-stretched schedule. Critical maintenance will not get done. Workers' experience burn-out, and increased tension in the work they do on a daily basis. Cuts that result in these problems are inconsistent with the goals of Lean Maintenance. Under these conditions, safety falls to second or third place.

It is obvious that, when a power cord is frayed, whatever brand of maintenance is followed, it should address the problem immediately. Clear and present danger is not the issue of these discussions. Even potential danger (one step down from clear and present danger), is usually managed pretty well by most organizations. In the environmental arena, comparable descriptions would be active discharge of a regulated material in excess of limits, and potential discharge of the material.

Lean Maintenance and Safety

The issue is where the maintenance is deferred, sometimes until there is an accident or a discharge of pollutants. Some companies then jump onto what they call a "lean" bandwagon and cut expenditures in many areas. We discussed the thought that it is unusual to cut your way into lean. Some of the areas cut might be related to safety or the environment, but the effects might not be felt for a few years.

Joanne Law, marketing head of the SIRF Lean Roundtable in Australia, spoke about a concept used there called Anorexic Maintenance. Anorexic maintenance is Lean maintenance taken too far. The cuts weaken the fabric of the maintenance effort. The result is a less-robust organization that is prone to accidents of all types. Consider the Alaskan Oil Pipeline, partially-owned and fully managed by BP.

In October 2006 Felicity Barringer of The New York Times reported "BP Exploration Alaska, the subsidiary of the international oil giant that operates the corroded transmission line from which more than 200,000 gallons of crude oil leaked, has been criticized and fined several different times, most recently in 2004 when state regulators fined the company more than $1.2 million. Now the division of the federal Department of Transportation responsible for pipeline safety is looking into the company's maintenance practices."

The article goes on to report interviews with maintenance workers. "In addition, one of the company's longtime employees, a mechanic and local union official who has participated in the spill cleanup, said in a telephone interview, that he and his colleagues had repeatedly warned their superiors that cutbacks in routine maintenance and inspection had increased the chances of accidents or spills.

In the interview, Marc Kovac, who is an official of the United Steelworkers union, which represents workers at the BP facility, said he had seen little change in BP's approach despite the warnings." Without judging the truth of the claim, or the motivation of the employees, it would not be surprising that a big oil company would defer maintenance activity in areas where the probability of a problem is low (in the eyes of accountants in Ohio or London), and the consequences of a failure could be handled (of course we can argue that point considering the fragility of the Artic environment, and certainly the PR disaster that followed the disclosure was probably more costly than the leak itself).

Chapter 6

The one point is that maintenance expenditures were certainly cut. Costly procedures such as smart pigging of the lines were put off. "Another question is whether the company postponed for too long a rigorous but disruptive internal inspection of the pipeline, known in industry jargon as smart pigging."

In the procedure, electronic monitors called smart pigs—successors to an earlier generation of cleaning devices that squealed as they ran through the pipe—are used to measure the thickness of a pipe's walls and detect defects. Mr. Beaudo and Mr. Kovac agreed that, since 1998, no such inspection had been performed on the line that leaked.

The work of setting up the pigging device is cumbersome, and its data are hard to analyze. The process also slows the movement of oil through the Trans-Alaska Pipeline.

BP's own 2003 plan for safe maintenance and management of its facilities, on file with the Alaska Department of Environmental Protection, says that "the interval between smart-pig runs is typically five years."

While BP's care of the pipeline does not seem to rise to the Anorexic maintenance level it is clearly on that end of the spectrum. However, the important discussion here, is to differentiate between Lean Maintenance and mere cost-cutting to make a profit number. Of course the Alaskan pipeline is over 30 years old, and it is an easy target because we have the advantage of knowing that it leaked.

The true test is what is happening with the reported million and a half miles of oil and gas pipelines in the USA, as of 2005 (according to the Bureau of Transportation Statistics. In the US, pipelines are monitored by the Department of Transportation). How are those pipelines being maintained? One way to look at the problem is to calculate the probability of a reportable level of leaks per mile, per year.

There are hundreds of companies responsible, so it is hard to look at the data. There are 10's of reported leaks in the past few years, or maybe less than 1 leak per 250,000 miles per year. So the very rough probability for the 800 miles of Alaskan pipeline might be around 0.003 per year, or 0.1 for the 30 years of operation. This estimate is not exact, but it might show that BP's history of 4 or 5 leaks might be 25–50 times higher than the whole field. Of course there will be intervening variables such as size, climate,. and other factors.

Lean Maintenance and Safety

All maintenance departments have some fat, and some fat can be attacked by cost cutting. Perhaps BP thought it was cutting fat. Just the act of forcing the maintenance department to do without, is not in itself an issue.

A sensitive company will have its ears trained on the workers in the field and their supervisors, and will listen intently for leading indicators of accelerated deterioration (such as in this example of minor leaks and incidents). Perhaps that is the lesson here. The company seemed to have passed the (invisible) point where the cuts start to impact the muscle and sinew of the effort. This cutting beyond the level of prudence will eventually show up in lower safety margins, and higher numbers of both incidents and of near misses.

Blindly cutting costs in maintenance without regard to the needs of the assets being maintained is completely consistent with increased leaks in a pipeline, increased breakdowns in a manufacturing plant, and increased safety problems in a refinery. The point of the exercise is to show that just cutting costs is not Lean. Too many cuts lead too often to defects and deterioration, which increases the probability of safety and health issues.

Many companies that are much less able to deal with problems than BP, are grappling with the crush of work in already downsized settings and "doing" Lean. The important question is, do you only know you have cut too much (and are too Lean) when it is too late? Unfortunately there are no straightforward metrics that show when you are in trouble.

One approach is to agree to a mission statement about Lean objectives and safety, termed fall protection. One such statement was developed by J. Nigel Ellis, PhD, PE, who is one of the leading experts in fall protection. He and his company and have integrated fall protection into Lean manufacturing facilities world-wide. His business, Ellis Fall Safety Solutions LLC provides the client with an approach to the problem:

"The goal of fall protection by a lean supplier of services is to protect against foreseeable fall and related hazards such that the fall distances and forces are minimized within technological feasibility and product application testing, to minimize or eliminate foreseeable harm caused by the specific fall protection system presentation and use."

CHAPTER

7

Lean Organizations and Maintenance Support

Benjamin Franklin says: Lost time is never found again.

Lean Business Systems are an important part of Lean Maintenance. You can't have Lean Maintenance with fat, or perhaps worse, non-existent support structures. The business system is the structure. It is the overhead (non-value-added activity) necessary to carry out maintenance activity. If maintenance activity was the product (if we were maintenance contractors), then any costs outside the work itself would be overhead. The goal is to reduce the non-value-added activity.

In *Lean Solutions,* Womack and Jones discuss bad processes. "How do we deal with this (bad processes)? Mostly we default to a 'bad people' analysis. The customer concludes that the provider is either an idiot or a crook the dealer decides that the customer is an overbearing incompetent, and the employees conclude that both the owner and the customer are a bad lot." This three-part analysis reflects exactly the attitudes in a plant with a broken maintenance management process. To quote Womack and Jones again "No one wins in a world with broken processes."

One of the European consultancies (Imants BVBA) was discussing some business process fundamentals as they related to Lean operations, in a recent article. Their discussion sheds some light on what to look for in sound business processes:

- A business process is a system of activities that creates value for customers (these are activities that the customer is willing to pay for).

Lean Organizations and Maintenance Support

- Processes are cross-departmental. Departments are functional towers of expertise, but processes cut across departments.
- Every process should be documented and fully understood by everyone participating in the process.
- To fully understand your processes, they should be mapped. Two main tools are used: flow-charts and process maps.
- Processes should be under statistical control.
- When processes are under statistical control, process capabilities can be calculated. These calculations can be seen as performance measures, and as bases for business improvement.

Most organizations are not designed. Usually they grow in response to the personalities of the people involved and the various stimuli from the environment. If there is a controversy, a new system will grow up to fix it. In the end, procedures at many organizations are cobbled together from years of reactionary growth. There is a whole field of business that looks at the process and optimizes it for the function being performed. This field is business process re-engineering.

Business Process Re-engineering is a way to remove fat from the staff organization. Even more than that, it is a way to remove wasted time from the daily activities of the maintenance workers. When some part of the maintenance value stream hiccups, or doesn't deliver what is expected—look deeply at the underlying process.

It is beyond the scope of this book to delve too deeply into the process design field. I would like to just skim the surface as a way of introduction. But first let's look at some examples of business processes gone crazy.

The work order for a city government had 43 steps from its issuance to its final filing. Several people touched the work order several times in its journey. Because it was such a pain in the neck, much of the work was being done informally, outside the work order process.

A defense contractor required maintenance technicians to source any part they needed, get quotes, and then submit the package to purchasing. Purchasing sourced the part themselves, and largely ignored the research by the technicians. To make it more interesting, if the variance in price between the technician and purchasing was greater than 10%, the whole package was returned.

A major manufacturer's maintenance engineer worked with a vendor to come up with a solution to a particularly intractable problem. The purchasing department didn't even let the vendor bid their solution, instead they went to a preferred vendor. The preferred vendor had lower prices and no local support organization.

These stories have broken business processes in common. Broken processes account for more fat than all other fat reasons combined. Of course, broken processes are usually much more difficult to see than are shown by the examples above. Such processes can be like getting an edict to scrap any inventory items not used in a year, difficulty returning parts that were not used on a job to the warehouse (easier to trash them or throw them into a rat hole somewhere), a lot of work being done casually without paperwork (because the paperwork is such a hassle), or people charging time and parts to any work order and not the correct work order.

One of the early unheralded leaders in the field was Abe Fineman of ICC Inc. He would look at the whole maintenance organization, track all the paperwork (computer flows), and optimize the flows, roles, and authority.

Paperwork Study including computer interfaces

One indicator of the business process is the paperwork and forms (on computer screens) used to control activity. More and more of these control systems are run by Enterprise software such as SAP. Analysis of the form flow, and especially the informal external systems, will tell the story of the underlying business processes.

Organizations use forms (either forms on computer or on paper) to control or initiate their repair activities. Over the years, these systems become more and more complex, and have control documents/systems added, and seemingly they are never removed. Forms are an excellent study to undertake for newer Maintenance Managers to help learn their operation in detail. The purposes of this study are:

1. Learn about all the forms, paperwork, and system activity
2. Investigate what forms/systems/procedures can be eliminated

3. Determine which systems/forms can be consolidated
4. Streamline queuing/waiting time
5. Plug any holes in the control of the operation

The study has two phases.
Phase I: Collection

Time: Choose a time that spans two accounting periods and includes one complete period. Usually 6 weeks, starting just before a new period is suitable.

Procedure: Collect copies of all forms, reports, charts, and any other paper (virtual or real) that passes through the maintenance operation (3-ring binders work well). The list includes repair orders, schedules, E-mails, purchase orders, reports, notes, little black book pages, matchbook covers, telephone logs, everything!

It's important to capture the little slips of paper because they usually indicate holes in the control systems. Some operations we've seen had excellent, computerized systems, which were ignored. The operation was actually managed from a black book in the foreman's pocket. For convenient access, these smaller documents can be taped or glued to 8 1/2 x 11, loose-leaf sheets.

As you collect these documents, you need to identify where they came from, numbers of copies, any ideas that you have about them, where they go, who signs them, who uses them, and any other interesting facts about them. These notes form the basis for the next phase.

Phase II: Analysis of Paperwork Study

Time: After the 6-week period is over and all the forms have been collected.

Procedure: Begin reviewing the documents and your notes. Keep in mind the five goals of the study. On large sheets (17 x 23 or larger, or using a computer program such as Visio), begin charting the movement of the major forms through the organization. Fineman uses columns to show people or stops that each form makes, and small symbols that are the shape of the form itself. Time is represented by the vertical axis of the larger sheet.

- What is the task or step
    - Provide a complete description
    - Identify predecessor and successor tasks
    - How is the work done?

- Resources
    - How many personnel are needed?
    - How long does it take to perform the task?
    - What special skill, specific experience, or judgment is needed
    - List systems or equipment needed
    - Any other facilities such as networks, fax, etc.

- Determine performance metrics
    - Quantify the amount of output
    - Can quality be measured?

Once complete, these charts will show graphically the complexity and route of the major forms. This system was used to uncover the 43 stops that the work order made in the first example.

Usually this study will result in reductions of up to 20 to 30% in the paper/form flow. The key is looking at all the forms and flows at one time.

In traditional Lean Manufacturing, rationalizing the business process is an implicit promise. In a list published in a White Paper by IFS (ifsworld.com) some of the following promises were discussed:

- Reduction in order processing errors (one of the keystones of the whole maintenance process is to make sure you are working on the right asset and doing the right job).
- Streamlining of customer service functions so that customers are no longer placed on hold (In our example, we want the customer to wait the shortest time possible, consistent with the priority of their job).
- Reduction of paperwork in office areas (there is a great deal of fat in the forms and paperwork—virtual and real) needed to run a typical maintenance department. A major effort is needed to streamline the number and complexity of the forms.

- Reduced staffing demands, allowing the same number of office staff to handle larger numbers of orders (There is so much to do that any reduction in staff workload is welcome. Let's let the planners plan, supervisors supervise and engineers engineer— so that they spend less of their time feeding and tending the business system).
- Reduction in turnover and the resulting costs of attrition (this sounds great, given the enormous costs of training and finding maintenance workers).
- Implementation of job standards and pre-employment profiling, ensuring the hiring of only above-average performers (imagine the benefit to the organization if everyone performed as well as the top 20%).

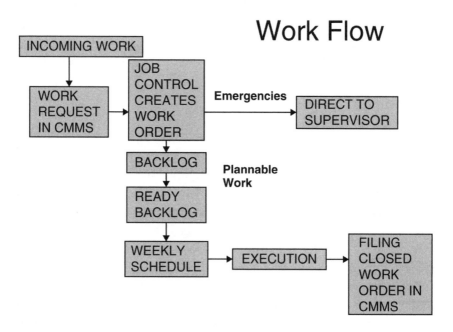

Work Flow

Effective Job control, the first key to lean maintenance

Take a look at the process used to accept work into maintenance. Job control has many names, but it is always the first place where work hits maintenance. Sometimes planners run this function. In small firms the supervisor takes all the calls and performs the function.

Chapter 7

Job control is primarily a customer-service function. Whoever runs it must be able to deal with irate customers.

Functions/activities of Job Control (managing incoming work is Lean)

- Time/date stamp incoming paperwork: All paperwork is captured and rationalized, no matter what the source (phone, E-mail, work request, radio). When the work order arrives it is time stamped. That is when the maintenance clock starts. All internal metrics are based on this time stamp.
- Triage then routes jobs to the appropriate player. This examination is the first peek at the nature of the work. Someone with maintenance knowledge reviews the job as soon as it comes in, to see if there is an emergency. Emergency jobs are routed directly to the supervisor (or directly to a worker in small departments). Plannable jobs are routed to the planner's queue for the planner to see.

Maintenance Technical Library (MTL)

Another structure in the maintenance department is the existence of a unified place for technical information. Information is one of the (briefly they are labor, parts, tools, equipment, PPE, custody, safe work, conditions, permission, information) essential elements of maintenance work. In the past, the MTL has been in a support role to the major items (such as planning). Over the next few years it will become one of the leading elements. It is the height of fat to arrive at the job with everything needed except the wiring diagram!

Information is stored in the maintenance technical library. The maintenance technical library is becoming increasingly virtual (meaning it exists wherever there is a connected computer). Even with an Internet- or Intranet-based technical library it is great to have a place where maintenance technicians can have access to a computer and a printer. Ideally, the MTL should be in or near the place that the Lean team's meet.

In such a space, maintenance technicians can use the local network (and local document management system), or the Internet, to look up equipment manuals, parts lists, assembly drawings, repair history jackets (in CMMS), and any information scanned in. The Library could also be a team room for Lean Maintenance projects, so it is the ideal place to house the studies, brainstorming lists, and the proposals. Last year's rejected projects could be one of the most useful places to look for savings.

One good unit to install in the MTL team room is a display cabinet for broken parts, bearings, and other small items. A display where you can pick up and handle burned bearings, fatigued brackets, and other worn items, and read details of the failure, can be a powerful teacher for new workers, plant people outside maintenance, and even maintenance old-timers. Clean them up and label them. You'll find people can't keep their hands off them (learning is happening!).

The MTL should also be the location of copies of plant drawings, site drawings, vendor catalogs, handbooks, engineering text books, etc. If you have computerized maintenance, stores or purchasing, CADD, or CAM, then access should be located in the MTL.

When the MTL is set up you will have: A ready reference for make versus buy decisions, repair histories, repair parts references with history, repair methods referrals, planning information sources, time standard development details, a data bank for continuous improvement efforts, and the maintenance improvement team headquarters.

The Internet is now everyone's Maintenance Technical Library

There are several ways in which the Internet is being used for maintenance information. The advantage is that the capabilities are available around the clock, 365 days a year (as long as you can get access to the Internet).

One last word—pick your battles

There are some battles that are longer-term and others that are shorter-term. What is needed is to recognize which type of battle you're fighting, and approach it accordingly.

Chapter 7

If you were trying to convince management of the need to redo their whole structure of something, that would be a big battle. It may be wiser to take a smaller project out of the big one and implement it, get the results, and then talk about those results. Then maybe you will get agreement for another project. Then you would be eating the big issue up one bite at a time.

You might see a need to change your whole business process, but that would be a hard sell. It's a hard sell because the people have certain powers that they don't want to lose, and they have certain influences that they equally don't want to give up. There is also a level of comfort and competence with the current system, no matter how arcane and inconvenient it is. Big changes will stir up big resistance. Small changes usually will flow below the radar, and can lead to the same result over time.

CHAPTER

8

Lean Maintenance Parts and Storerooms

A global inventory reduction expert, Phillip Slater says "Applying lean principles to maintenance storerooms not only enables more efficient operations but also assists in the elimination of excess and unnecessary spares holdings, resulting in higher stock turns and greater availability." Let's start off this subject by defining some terms. Lean (when referring to parts and the warehouse) means having exactly the parts you need and not one item more, nor one item less. Lean also means that when we use the last item and need a new one, the truck pulls up and hands us the exact part we need in the quantity that we need. In this situation, Just in Time and Lean are siblings. This condition is the goal of a lean inventory.

Of course, redesign that eliminates the part usage in the first place is the leanest solution. For the purposes of this discussion we'll say that redesign is not an option, and that we will need the part after all. We'll discuss redesign to achieve leanness in another section.

All kinds of companies work at cutting their maintenance inventories. The question is, "what is the 'right' level of inventory," or "what is the 'Lean' level of inventory".

One of the leading thinkers in the inventory area is Ronald M. Schroder. He discussed two truisms in relation to the stocking policy. The first one was "No amount of inventory reduction can save enough to pay the cost of getting caught short on a highly-critical part." Going hand-in-hand with that thought is the statement "No amount of inventory can guarantee that there will be no shortages (stock outs)." This thought seems like a contradiction of the first, but it actually stimulates a creative tension between the two schools.

The parts room or warehouse is one of the primary areas of fatness (or lack of leanness) in the whole maintenance organization. To make up for stockroom inadequacies you'll find all types of anti-lean practices such as rat-holes with hidden part inventories, and tool box inventories. You'll also find trades people rebuilding and repairing parts not designed for rebuild because they don't have the parts needed for a critical repair (and cannot reliably get them).

These expediencies do not even mention the time lost waiting for parts supposed to be in stock and not on hand, or parts that are in the warehouse but cannot be found. On the other side of the spectrum we find parts being discarded because either it is too difficult to return them, or there are too few stockroom employees to figure what the parts are and where they belong.

Inadequate stocking systems and policies account for the majority of lost time, majority of downtime, and by far the majority of the headaches suffered in maintenance. Many people complain that purchasing or finance owns the inventory, yet the maintenance department is held accountable for the results, such as downtime and MTTR (Mean Time to Repair).

In the name of 'Lean,' inventories have been decimated. To be more personal, perhaps your inventory has suffered this fate. The question is, is this the best way to proceed for the leanest overall organization? There is no doubt that most maintenance inventories have fat. In fact the warehouses run by maintenance folks might just be the worst of the bunch. What is the worst choice for the long-term company profit, having too much inventory or having too little inventory? We must look deeply into the definition of Lean maintenance to find the answer.

With my apologies to the accounting profession, and to conventional wisdom in cost-cutting circles it is clearly more expensive to run out of critical items and not have the items (especially due to a inventory reduction effort, where a part that was owned was discarded), than it is to over stock them. Keeping a supply of parts needs the following:

Lean Maintenance Parts and Storerooms

1. A building with adequate storage and limited access, good lighting, and a controlled environment. If there is no crewing at night, an easy fool-proof way must be worked out for access, part look-up, and part sign-out.

2. A requirement that all parts removed are recorded on WOs or equivalent document. All parts used are related to an asset for which they are used. Some means for recording the quantity used, the date, where used, and approved substitutions. Waiting time is minimized by preprinting pick lists that can be picked off-line and made ready for the tradesperson. The stores is crewed so that there are enough hands during peak times at the beginnings of shifts, and around breaks and meals.

3. All parts must be received, which mandates basic receiving checks such as: Part number compared to what was ordered, physically checked (is it the right part? Damaged?) and counted.

4. Parts are assigned locations, and shelved. All part locations are logged on the inventory system for quick retrieval. Packaging, strapping, and plastic wrap are removed, but boxed items are left in boxes for protection.

5. Periodic physical inventories are taken to verify quantity and location. Cycle counts usually are satisfactory. Bin locations are periodically audited to insure that only the right parts are stored in each bin and that any empty bins have parts on order.

6. Each part has a master part record with price history, usage history, and cross-reference numbers (various manufacturers' numbers may exist for the same part). Other fields include the ways that the parts are found and/or identified. The fields would include the names of the manufacturer and the vendor, generic type (bearing, seal, belt, etc), standardized short and longer descriptions, where used (what assets the part is used on), when used last, category and degree of criticality.

7. Periodically, parts are shopped and vendors evaluated. The system used can easily generate parts catalogs, so periodic shopping for higher-volume items is very easy. Different channels of purchase are investigated if warranted by the volume.

8. Periodic reviews are conducted of applications and specifications of parts ordered. Vendors are used to suggest more robust or cheaper substitutions.

9. Periodically, part usage and lead-time are reviewed to adjust re-order points and economical order quantities. The real power of computerization lies in its ability to capture and analyze usage data and apply preset formulas to determine Economical Order Quantities with little additional effort. The computer system can apply the check for minimum stock level every time a part is requested, ordered, or received. When the parts are ordered when requested (instead of being held for batching) then stock-outs can be reduced to the level that you set. Inventory levels can be adjusted up or down by allowing more or less stock-out conditions. Once everything is settled down, fine-tuning for seasonal variations, and for the age of the equipment, can bring inventories into line.

10. Parts are divided into classes for different treatment. ABC classes of parts are set up through analysis of yearly dollar volumes. Categories for critical parts such as "Insurance policy spares" (parts that could shut down a section, process or the whole plant) and "critical parts" (parts that can shut down a whole machine) are very important.

11. Inventory that hasn't been used should be tracked. The system will print lists of parts that have not been used in 6 months, 1 year, or 2 years. These parts can be investigated (to see if they are hard to get insurance' stock). If they are available from outside sources you can try to sell or trade them for usable stock. Lean projects can involve disposal or sharing of these slow items.

12. Information is available so that questions can be answered.

13. Parts for assets that have been retired are reviewed for use elsewhere or disposed of. Outside information such as asset retirement, and changes in asset make-up, usually have to be manually factored in. The system has where-used as a data element in the part Master files, and it will have the ability to identify all parts used on the retiring unit for review.

JIT Inventory

There is a JIT (Just in Time) opportunity in a typical maintenance inventory. Many people ignorantly de-stock the warehouse in a philosophical effort to line up what is happening on the shop floor, with a minimum of raw materials and work in process. Such actions have been shown to be ill advised. The issue is that the organizations that supply M&O parts, the maintenance supply chain, are also moving more toward JIT for their own plants.

Like a good JIT student, the whole supply chain is operating on a pull basis (parts are being made when they are ordered). Distributors who used to stand between the manufacturers and the customers and buffer the demands, now only want to handle spares that move. So on anything even vaguely exotic, the supply chain is empty, and demand is used to pull the part from the OEM to you. There is nothing wrong with this concept except that the lead times for most spares have increased. Our Predictive maintenance (both the technology and the mental approach) has got to be a lot better if we expect it to alert us when a part is going to be needed, well before it is needed.

JIT Opportunity

There is an opportunity with parts used for PM, such as things like filters, belts, seals, and anything that is replaced on a schedule. We can order these items a year in advance, for drop off a day before the scheduled PM task. If we do the PM monthly, we can get a monthly release and delivery.

This opportunity is also useful for consumables, where you can predict the usage. One large plant had the janitorial supplies delivered right to the various janitors' closets every month. Only a small safety stock was needed in the warehouse.

Make versus buy as a route to Lean

Sometimes the solution is to have the part made, rather than buy it from an OEM. Maybe you can have the part made better or cheaper than the OEM is making it. When the OEM goes out of business, or stops supporting your machines, you might be trapped into this choice against your will. A lot of companies spend some brainpower on this issue, and it might save a great deal of money. Of course, the OEM has specific expertise that you might have a problem duplicating. There is significant liability because some of the specifications are invisible and may be critical.

A better approach, which we will discuss at length, is to figure out why you're using that part. This process might involve everything from a Root Cause Analysis project to a lucky guess. Then you will either fix the problem or build a more robust part, which would lengthen the mean time between failures. Fewer parts will then be used. Money to buy the part will be saved, plus the labor to remove and replace the part, and the downtime from both the breakdown event and the time to replace it. That's Lean!

Rebuilding used components

Is there life left in a component? This is an important Lean decision. If there is life left, then a rebuild might be justifiable. But be careful, it can be a trap. The life of the rebuilt component has to equal or exceed the life of a new item, or you may save dollars above the water-line to pay hundreds of dollars below the water-line. Unless the economics is clearly negative scrapping parts that have good life left, or are rebuildable, is Fat.

A significant area of interest is in re-using and re-building used parts. In some industries, where assets suffer significant use and wear out, such as in mining or construction, used parts or rebuilding used parts is sometimes well established, so might be your only choice. Such customs may also have significant advantages:

- In these industries there is significant incentive to not let the asset go all the way to catastrophic breakdown. Core damage (complete destruction of the re-buildable part) is less likely.
- The re-build can be scheduled and there usually is a wide window for when the work has to be done. This ability to schedule can even out the workload in the shop, and provide useful employment during down periods.
- Tools used in the re-build can be made available on a scheduled basis to reduce conflicts.
- Manufacturer's revisions, enhancements, and improvements, can be incorporated more easily. Many re-build departments boast that their re-builds exceed the life of the OEM parts. In general the re-builds in controlled environments by specialists are always better than the same re-builds `on the floor.

Lean Maintenance Parts and Storerooms

Warrantee for parts

It is Lean to have vendors pay you when their parts don't last for their guaranteed life. Most large or expensive parts have warrantees. Improvements or modifications to the warrantees can sometimes be negotiated into the purchase contract. One such common negotiation is to change the warrantee start to the date in service rather than the date received.

In most warehouses, money is left on the table (not even asked for) for unclaimed parts warranties. The question is how many companies collect all the warranty money that they are entitled to?

Manufacturers tell us that they reserve 1%–2% of the sales price of the parts or equipment to pay for claims. Although most plants recover some warranty money from new equipment that fails, there are plants that recover no warranty money for defective parts. Warranty is money owed to you from the manufacturer of your parts, even if it is paid in spare parts. As such it should be managed as intensively as any other receivable. In an article in the July 2004 Maintenance Technology magazine, Joe Mikes outlines the information that must be collected for any claim to be pursued, as follows:

1. What was the unique ID number of the asset the part was on (or in) when it failed?
2. What was the date and where was the part bought?
3. What were the dates and/or meter readings when the part(s) were put into service.
4. Provide documents of proof of purchase.
5. What are the warranty guidelines for that part (6000KM, 3 months etc)?
6. If the part is also a major asset such as an engine, provide proof that specified PM was done (such as oil changes).
7. Where there is a substantial amount of money involved, photographs of the broken part might be useful.

Where does this data come from? The primary source of data is the work order system of the CMMS. The work order has the unique ID number of the asset and has the date when the part was put into service. Another work order has the information about the failure.

Chapter 8

The warranty guidelines should be filed in the part master file that has all the fixed information for each part. The purchasing department has proof of purchase, the reference number (purchase order number) is in the part transaction record.

If you have to reconstruct PMs for the unit, between the date in service of the part and the breakdown, you would look to the work order system. The search entry would be for all PMs between the date in service (of the part) and the date of breakdown (again of the part) for the unique unit number. A photograph is a useful addition to the documentation package for expensive parts, or for parts that failed in unusual ways.

This package is to be sent to the vendor with an invoice for the labor and the replacement part. Usually a replacement part is adequate reimbursement for the failed part, though there may be a large discussion about the labor charge (and paying the labor costs at all).

Lean Maintenance and the Work Order System

The work order system is one of the most confusing areas for Lean maintenance proponents. On one hand we have the indisputable value of accurate data when investigating other Lean projects. On the other hand is the high cost of collecting the data in the first place. For example, the Municipality of Dubai in the UAE has over 600 buildings (and growing). They have 40 people doing data entry alone. The Lean question is, can this amount of effort be justified? What if we made the 40 people redundant (or reassigned them to other tasks within the municipality, which is more in keeping with Lean rules), and went back to paper work orders and verbal instructions?

One observable phenomenon is that, when work order systems are installed and followed, the discipline seems to improve the productivity of the people in the trades. The instruction that all work must be reported seems to have a positive effect. There are surely psychological reasons for this improvement but it seems that the systems often pay for themselves in higher productivity. Such an advance in itself is not enough to recommend installing a CMMS. We have to look for other advantages.

Doing the right job right is Lean

Out of all the approaches to waste reduction and Lean Maintenance, at the top of every list, is doing the right job right, the first time. Two issues with different causes are described here. There is nothing fatter then having to return to a job that was done incorrectly. There is also nothing fatter than doing the right job to the wrong asset (or doing the wrong job to the right asset). These are all quality issues but from different parents. We will deal with the first (jobs done incorrectly) in the Chapter on quality.

Doing the right job to the wrong asset (or the wrong job to the right asset) indicates a breakdown in communications, in memory, or in nomenclature (assets not having fixed names). These are serous issues that, when they are ignored, could result in unintentional and unnecessary accidents, or production outages. In fields with high intrinsic hazards such as aircraft, chemical plants, hospitals, nuclear power plants, or oil refineries, extreme measures are taken to avoid these kinds of mistakes.

Probably the most painful place where such errors occur with alarming regularity is the maintenance and repair shop known as the Hospital. The machines being repaired are humans, and the Evening News frequently reports mishaps of such errors type in hospitals. Mistakes in a hospital usually only affect one or a few people, but mistakes in a refinery could affect dozens or even hundreds of workers.

Instructions to repair or replace a specific asset are written, verbal, or both. Each mode has unique issues. When giving instructions verbally, there is always the possibility that the instructions were not heard, mis-heard or forgotten (and then replaced with incorrect information). When there is no record of a conversation it is hard to pin down where the information breakdown occurred.

With verbal instructions there are various places where the communication takes place including face-to-face, over the phone and over the radio. One of the big variables is what was being interrupted to give the instructions. The situation is much different if the communication is in the shop, in relative quiet, before the day has begun, than if you are interrupting work on a complex job with a radio call.

When giving instructions in writing (generally by work order) there is always the chance that the reader can't read well enough to understand the instructions, can't interpret what's written, or that there is some ambiguity in the asset identification or job description fields. Hand-written work orders are susceptible to being unreadable because of dirt, or bad handwriting.

A sound and efficient approach to communications is essential to minimize the number of mistakes. The approach that uses the least resources and delivers the most accurate results is the leanest. There are many advantages to computerized work orders in both the short term (reporting discipline and communications) and long term. In the long term, computerized Work Orders are essential for data analysis. The issue is that Work Orders are Lean as long as you get the information cheaply and accurately, develop a system of data management and training, and then use the data.

So, if written work orders are Lean when compared with verbal (assuming the rules are met) what about the use of CMMS? If fact, the way many CMMSs (if not most) are used, is not Lean at all because the data is not complete, not accurate, or it is unsound. Collecting incomplete data is Fat because the value of the data is in its use to improve the decisions that are made. With incomplete data, decisions are not improved. Incomplete data works against any other Lean maintenance initiatives because bad data may result in erroneous conclusions. The results might result in pursuit of a Lean project that is really fatter than what was done before.

Without going too deeply over the ground with a work dedicated to the CMMS, there are a few things that would make the current crop of systems more useful. If we had a metric it might be something like the unit of usefulness of the system, per hour of input. In some circumstances we can treble our usefulness with little or no increase in hours expended. In another section we will discuss the use of advanced technology for data acquisition, such as wireless PDAs, which will change the picture.

The problems connected with collecting and using maintenance information were first faced by the trucking industry in 1968. With a grant from then Union 76, and under the banner of the American Trucking Association, teams met to discuss computer systems and data entry issues. Keep in mind that the first known CMMS was called MIDEC and was designed for Mobile Oil around 1965.

The conference met many times during those first years and developed a scheme that was called the VMRS (Vehicle Maintenance Reporting System). VMRS has been in constant and wide use in the trucking industry since then. This system simply and exhaustively codified everything that could be done to a truck, a trailer, a construction unit, or any other piece of mobile equipment.

Chapter 9

Every item was classified into a series of numeric codes (this was early in the computer field when numbers were easier to deal with than letters and words). With repairs, the highest level was system codes, which were two-digit codes for the system on the truck that was being worked on, such as code 13 for brakes, or 26 for the automatic transmission. The next level (2 digits) was the assembly code, such as 001 for front brakes. The lowest level (3 more digits) was the part itself.

In addition to the number breakdown there was a work-accomplished code to identify such work as repair, remove and replace, inspect, lubricate, etc. A combination of the two codes tells exactly what was worked on and what was done. Another code asked the mechanic to identify the major mode of failure such as corrosion, fatigue, etc. This information was particularly useful for forensic failure analysis, especially when there were a few years of data on many similar units.

Starting in the late 1970s this coding enabled the computer system to develop reports for component analysis (how do these brake pads compare to those), rework (when the same system is worked on a second time within 4000 miles), and component life (elapsed mileage between failures on any component such as steering or cranking). Typical fleets had hundreds or thousands of similar units. The large populations enabled statistics to be used. Statistical data analysis of fleet breakdown data was on its way.

Try this thought experiment. With your current system and your current data collection standards, think about the ease of generating a report on the average elapsed time between failures of a hydraulic pump (say a component in an injection molding press). How hard would it then be to compare that statistic with those for similar pumps in other molding machines? If that seems overwhelming, think about why that is so.

One shortcoming of current implementations is the inability to use the power of the computer system for such a level of analysis. Some 90% or greater of the CMMS implementations do not have the data coded into the system to perform this level of detailed analysis with the computer. Some, to a good deal, of human intervention would be needed to prepare reports like those discussed above. Lean implementations of CMMS lean (sorry) on the computer to do the heavy lifting needed for analysis, and the data must be entered in a way that the computer can analyze.

If we expect the CMMS to be a tool in the Lean tool belt it must be made useful. To be more useful the computer needs to deal with completeness, accuracy, and soundness.

How many of the following list of symptoms do you have? (The more you have the worse off you are)

- Work (labor) is charged to the closest or most convenient work order, rather than the correct one.
- Parts are charged to the closest or most convenient work order rather than the correct one.
- Where estimates are printed on the work order, the actual time required for the work is always the same as the estimated time.
- Time is rounded to the nearest hour or time is picked out of the air.
- Work orders are left open for a week or a month, and they accumulate all the work that comes up (called standing orders).
- Anything under 4 (or 2) hours is not recorded.
- If the person is given 4 work orders, 2 hours are charged for each order.
- People fill out the work orders at the end of the day and 'remember' what happened (without notes).
- Jobs are done where there is no work order at all.
- Work is done but the work order turned in has no records of hours or materials and no explanation, or just says 'OK' or 'Fixed'.
- What was actually done is not clearly stated.
- What was actually found by the mechanic is not clearly stated.
- The whole day is not fully accounted for (such as time in training, meetings, etc.)
- Work orders get lost and no one remembers what was done or when.

Any combination of these symptoms could undermine the integrity of any database. Our focus in this work is not on the details of the data collection, but rather on the use of the data after it is entered into the computer.

First we have to establish that all or most of the hours paid for are recorded in the system. You can do this by simply asking your accounting department to run a report of the payroll hours for the maintenance workforce by type for the last month. Using the CMMS, run an all-hours report for the same period. After juggling around the hours for people on leave, check whether the two numbers are pretty close? They should be within 5% of each other. Note: you may have to deal with what happens to hours spent on unfinished jobs (in some systems they are not reported until the job is complete). In some shops the work orders are held by the mechanics until the work is completed. This procedure will skew the numbers for the short term.

Do people agree that the data going into the system is accurate? This is the next hurdle. If we've established that the information is complete we have to confirm that it is accurate. Is the right work order being charged? Generally a quick check of the raw work orders will indicate quickly whether the data is reasonable. A more-detailed audit and spot check might show some areas for improvement.

Three categories of problems with work order data

Incomplete: There were jobs that were done where no work order was taken out, and the jobs therefore are not in the system. Sometimes a contractor is used and the work was not recorded against that asset number (so it is not in the database).

Unsound: Hours are charged to the nearest work order, not the correct work order.

Inaccurate: Work Orders contain one or more of the following: wrong part numbers, wrong asset numbers, wrong hours, wrong contractor information, and wrong workers.

Once you've established that the data set is complete and accurate, you need to check whether the data is useful or sound. Take a handful of work orders and see if these simple questions are answered (without being Sherlock Holmes). If these questions are answered satisfactorily, you have a good shot at using the CMMS database for your Lean Projects. Here are a few data elements that should be collected.

- What did the mechanic find when he/she got there?
- What did he/she do about what was found?
- How long did it take to repair or to return the asset to service? Breaks are generally included, but lunch is not.
- What materials were used, listed by part number and quantity?
- What other resources or services were used?
- If there was planning, are there actual numbers (as far as materials and labor) to compare with the plan?

Job control as a source of Lean Maintenance

One Lean idea was pioneered by Xerox over 40 years ago. Xerox was the first company to place complex copy machines in offices. At the time, these machines were the most sophisticated, complex, and temperamental ever placed in the public's hands. Xerox generally placed the machines with full service leases. If the machine was sold instead of leased, the buyer would always take a full service contract (again because of the complexity and lack of reliability).

The Xerox service staff found that it was responding to large numbers of service calls for which there was no real service problem, or if there was a real problem they were not prepared with the correct tools or spare parts. After grappling with this situation, and trying various strategies, they came up with a procedure called 15 questions.

The issue was that Xerox wanted to ask a series of 15 simple questions to find out what was going on without angering the customer. They also trained a person in the customer's office to be the primary contact, and called that person the key operator (because that person was given the key). The key operator was taught all the questions and shown how to get the answers.

The questions were designed to eliminate reporting of non-service related problems and to direct the service dispatcher toward making a more intelligent decision about what the problem was likely to be. One of the most common problems was the machine being out of paper. The question was phrased "Please check what kind of paper is in the machine." Of course the key operator would check and find no paper, and the service call would end there. Another question was designed to check whether the machine was plugged in. The question there was "Please verify which outlet the machine is plugged into."

The 15-questions strategy reduced service calls by 15%, increased customer satisfaction, and dramatically improved the FTRR (First Time Repair Rate). This strategy required skilled (and slightly more expensive) call center employees who sorted out the jobs and filtered out the ones that could safely be dealt with by the customer. The skill applied at the front end resulted in more productivity on the back end. Amazingly the customers loved it.

The Lean question for your company is how well do you investigate incoming work? Do you investigate it at all? Are there any opportunities for front-end filtering?

CHAPTER

10

Lean and the Use of the CMMS to Uncover Waste

Before we talk about using Computerized Maintenance Management Systems to cut waste, I want to look briefly at enterprise level systems. These systems are used to integrate all or most business functions into one large system that shares data. The same employee master file is used for payroll and for the CMMS section. Some of the largest vendors include SAP and Oracle.

This system offers great potential advantages for the purposes of Lean Maintenance. Some of the problems outlined in the previous chapter become less problematic when data is shared. For example, if information about work order hours is fed to the payroll program, that data will be more accurate (it had better be). These systems cost $1,000,000s with consultation, so that top managers are looking closely at the results. With such costs, the importance of clean and accurate data is pushed by all levels of management right up to the president.

With significant savings being offered to other business users throughout the company by enterprise level systems, adoption must be a decision at the corporate level. One caveat is that to get the benefits, significant money must be invested in training in the use of the maintenance section of the system. However, many companies seem to run out of funds right before they get to the maintenance implementation, which effectively leaves the maintenance folks to do it themselves.

From a Lean point of view, how much time and training must be invested to master the systems? In a typical CMMS, a hundred hours (some of this should be on the job) are needed for a basic knowledge of the system. Years, and multiple training sessions are needed to get really comfortable and effective at using the systems to ferret out waste. The enterprise systems mentioned above can take two or three times the normal amount of training for users to get comfortable. That investment is not always made.

The CMMS (Computerized Maintenance Management System) is the repository for all the data generated by day to day maintenance operations. In so far as the data is sound, complete, and accurate, it provides a great view into all kinds of waste. Some of the waste that can be seen with the data cannot be seen any other way.

As discussed in the previous chapter the issue is that the data is frequently incomplete, unsound, and inaccurate. The CMMS is then worse than nothing, because it might indicate actions that are contra-indicated, given the numbers from good data.

This chapter assumes that you have adequately addressed the issues of the data and you have a level of confidence that the data are good. Once that hurdle is crossed we can look at some of the kinds of analysis that would reduce resource use.

Determine that PMs are due soon, when a unit shows up for service: let's start with an extremely easy query that can be answered by a CMMS (and that should always be asked before someone repairs a breakdown or performs corrective maintenance). The question is "Is this unit due, or almost due, for any level of PM Service?" Studies show that a good deal of maintenance labor is consumed by travel, getting an asset locked out, getting parts and tools, and getting the job assignment. By combining the repair with any due PM, you automatically increase the productivity of the maintenance worker.

Component Life: All machines are made up of components. These components include power (motors) and power transmission (shafts, belts, bearings, couplings, etc.), protective devices (fuses, P/T relief valves), controls (PLCs, controllers, I/O modules), conveying systems (feeders, conveyers, piping, pumps, etc.) and working parts (jaws, blades, rollers, etc). Each of these components has failure modes and failure frequencies.

Lean and the Use of the CMMS to Uncover Waste

The component life analysis involves looking first at the failure frequency of each type of component to see if there are any clear trends (either toward longer or shorter MTBF). The second thing is to look toward the specific failure modes and see if there is anything to uncover there. If you have different component makes and models in similar service, then you can compare the two. There are many variables, so it is best to go slowly.

In an older company, senior maintenance professionals and engineers had decades of experience and even longer traditions to bring to bear, on the problem of choosing the best components. But in the last 20 years as our plants got more complex and the departments were downsized, senior maintenance professional time became more valuable. Demographic trends exacerbate the problems, with the high numbers of retirements from 2000 to 2015. These masters are now focused on keeping the plant running, and not on any deep thoughts. We have sacrificed the ability to use the hard-earned experience of these personnel for component-buying decisions.

Even without the push toward cheap maintenance there is some doubt that today, anyone can see enough and know enough to make these decisions in our highly-complex, high- technology plants. But with all that said it is still well known that significant Leanness can often be derived from finding the best component systems and sticking to them. The data to make these decisions is partially in the CMMS files. Component Life Analysis in what ever form you pursue it, will help bring out some of this essential data in formats that can make a difference in your decisions.

Rebuild or buy a new one: One area ripe for the harvest is using hard data to help make the rebuild-or-replace decision. Many companies scrap components when significant life is still left. Other companies reuse items that would be better discarded. Some cores (the item to be rebuilt) have specific lives, depending on the service. Retreading tires is a good example. A tire casing can be retreaded a few times, particularly if the truck is used in light service (such as furniture moving). A steel hauler should be more careful (and won't get as many retreads out of each casing).

Chapter 10

When you have significant history you can determine if the life of the rebuilt component is long enough to justify the rebuild. One rule of thumb was that the rebuilt unit should last greater than three-quarters of the life, and cost less than half the price of the new one. Of course the labor cost, downtime cost, safety implications, and other consequences will enter into the decision.

It's a happy day when you can rebuild a pump and get better life out of it than out of a new one. The Leanest way to rebuild is to improve beyond OEM specifications. It takes a good deal of engineering expertise, but it can be worth it. Sometimes you can borrow expertise from the bearings vendors, lubricant vendors, and others who can make money out of this strategy.

Repair or replace: Related to the above analysis, and similar in the use of the data, is the decision to retire a machine and buy a new one, or repair the old one. It's an old story; the pump breaks again, for the fourth time this year. Should it be rebuilt again, or should we discard it and install a new pump? Significant mathematical work has been done in this area by Professor A.K.S. Jardine currently at the University of Toronto. In every case the success of the analysis is dependent on the quality of the data.

During a Lean discussion at a saw mill, the maintenance workers made a list of the machines that required a lot of their time. Some of the machines (in particular a crane that selected the logs before the barker machine that removes the tree bark) broke due to operator errors. The data showed that the costs incurred (especially when both above- and below-the-waterline costs were considered), were excessive. It seemed that a new crane could have been purchased every year for what was being spent to repair the old one! Of course, when it turned out to be an operator error it could be argued that the operators should be trained, disciplined, or otherwise sorted out.

When we discussed the problem with operations we found that the failure mode resulted from taking short cuts to make up lost production (usually due to minor jam-ups that were unrelated to the effect). It was found that, if the operator was really careful, the crane could take the load.

Lean and the Use of the CMMS to Uncover Waste

We could repair the crane, or we could replace it with one that is more robust. The Japanese take a different view. The Japanese would mistake-proof the system. In TPS (the Toyota Production System, the home of Lean manufacturing) there should not be an operational mode that will hurt the equipment. They call this process poke-a-yoke or mistake proofing. Perhaps on the next rebuild, the saw mill should reengineer the system so that the operator either cannot hurt the equipment, by limiting its range, or by beefing up the equipment to permit the heavier strain induced by the short-cut operational mode. We would choose to beef up the equipment because the "wrong" operational mode has some distinct production advantages. We might even 'thank' the operators for discovering a faster way to operate the equipment.

Making good repair-or-replace decisions lies at the core of the expertise needed to run a Lean Maintenance operation. Given that much of the forward-looking data is going to be unknown at the time of the decision, it behooves us to quantify everything we can, now. Good decisions require real data. Real data can be found in the CMMS.

Absolute money spent: Sometimes the amount of money spent on an asset in absolute terms will be the indicator that the asset is ripe for replacement. This is a "closing the door after the horse has left the barn" type of reaction. There is a school of thought that says, if components start breaking, then it is possible that the asset has reached the end of its life-cycle. At the end of the life-cycle, all kinds of things will start to break.

With enough data you can sometimes detect when an asset is going to turn the corner and things will start breaking, like popcorn in a microwave. The idea is to retire the asset as soon as possible, when you realize you've hit the start of the breakdown life-cycle. Lean says that at the very least you should do the investigation before making any major investment in an asset. This analysis is best done before spending major money on an older asset.

Where to direct the engineering talent and the RCM folks: Any company that is lucky enough to have reliability experts or maintenance engineers faces a dilemma. Given all the potential problems to focus scarce resources upon, what problem would give the best bang for the buck? There are always more opportunities than there are hours. There are a few major ways to decide what to work on. We could focus on the most critical assets, the most dangerous failures, the most expensive failures, the failures that take up the most maintenance time, or the failures that are most disruptive to the manufacturing process. Of course, these categories will overlap.

Each of these approaches requires different data sets. The criticality of a machine, other asset, or process, is a decision to be made by people who are knowledgeable about the plant and process. Most CMMS have a field for criticality, which would aid in sorting out the potential areas to focus on.

Most expensive failures are the easiest analysis performed by a CMMS, as long as the hours, part costs, and details about contractors and outside services are complete and correct. The query would just sort all the repairs for the time period and display them in descending order of cost. The engineer could then just pick from the first page of the list. Lists of failures that consume the most maintenance labor would be developed in the same way. Just write a query that displays the jobs by hours consumed, for the period in descending order. The first few entries would be the most expensive.

The most disruptive failures would use the CMMS data about downtime (when that information is collected), added to a series of interviews with people in operations at various levels. The interviews would capture shorter and potentially more disruptive events, than would the system. The level of disruption might not always be linked to the length of the breakdown.

The most dangerous failure modes would have to be gathered from an interview/brainstorming process. The interviews would include safety professionals, maintenance and operations and process experts.

Using the system to compare like units and to find the machines/assets that are exceptional (both good and bad). The power of the computer is that it can be used to quickly and efficiently compare large data-sets. One of the questions that can be answered by this kind of comparison is which unit is the best of the bunch.

Lean and the Use of the CMMS to Uncover Waste

The plant might be running hundreds of pumps. The 80/20 rule says that 80% of the problems are coming from 20% or less of the pump population. Much of good management practice is to determine the exceptions and manage them.

Use the system to find those bad actors (the pumps giving you all the headaches). Distinguish the bad actors, and determine which variables are responsible for their badness. Generally, figuring out what is going wrong and fixing it, will give you the biggest return on your time investment of any intervention. You also might want to find the 20% good actors as well, so as to replicate the conditions that produced their goodness. We have a natural tendency to be attracted to studying the bad actors, the units always breaking. It is useful to also spend some time with the good actors, the ones that don't give us any problems.

Finding the mechanics that need training or other intervention: It will not surprise anyone in the field, to know that many maintenance problems are really people problems. Repetitive breakdowns might be the result of a lack of knowledge. Low productivity might be the result of a hidden health problem. The data is there in the files, just waiting for a supervisor to take a look. The system should not be used for discipline. But it is a good idea to review production levels, rework, time on the job, and other indices, for potential issues. All this data review would be tempered by the supervisor's own sense of what is happening.

CHAPTER

11

Enabling Technology for Lean Maintenance

This chapter really needs to be rewritten every few months to reflect changes in the state of the art, and take account of innovations in such widely disparate fields as tele-communications, medicine, aviation, computer science, and AI, Internet II, mobile computing, imaging, pattern recognition, nano technology. All these developments and hundreds of others can have an impact on the delivery of maintenance service. Anything that has an impact on maintenance service can be twisted around to be used for Lean purposes.

Probably the biggest impact has come from Moore's Law which states that computer power will double or cost will halve, every 20–24 months. Having gigahertz of processor power and terabytes of storage at reasonable prices makes possible things that were impossible a short time ago. These capabilities are being exploited by vendors, with new features and capabilities.

For lovers of new technology this current period has been great to work in. As in any field touched by technology, entire industries have emerged and disappeared. We can certainly look forward to increasingly sophisticated gadgets, tools, software and systems, to help us to manage maintenance. Lean maintenance uses new tools but does not depend on them. W.E. Deming issued a warning that gadgets will not fix the problems with quality.

The same thing is true for Lean Maintenance. Gadgets are great, but buying them will not make thee lean! Lean is a mental attitude that is supported by whatever tools you can get your hands on and use. With the right attitude and training, a pencil is more powerful than a supercomputer for Lean Maintenance.

The focus of this section is to look at the ways that disruptive technology can be used, and the wide categories of technology to look out for. Technology firms are working overtime to provide technology solutions to (they hope) improve maintenance. It is tough to get the attention of the maintenance audience, so these firms are very competitive with claims and anecdotes. Any time that worker hours can be replaced by a device that is reasonably priced, we win a small Lean battle. Obviously the goal is to be sure that the humans on your crew are devoted to the highest skill activities that they are trained and competent to do.

A few years ago it was impossible to have economical access to the extremely large databases of documents, manuals, and other maintenance information. Today, desk-top systems are shipped with terabytes of storage. Each new jump in power and storage capacity allows new levels of operation. Enabling technology comes in several forms, and each form has products that are identified with it.

One such disruptive developing technology is the use of AI (artificial intelligence) to aid troubleshooting. Although it is common in large machines (such as turbines), and high volume machines (automobiles), AI is an uncommon application for general maintenance purposes. Dr. M.M. Nadakatti, an Assistant Professor at the Gogto Institute of Technology in India, has been working in this field for a number of years (he wrote his doctoral dissertation on the topic), and has developed software to support the process. The issue is to capture the knowledge and internal thought processes of the subject matter expert, and translate it into computerized rules. This kind of progress is essential as the senior people retire, so his work, summarized below, comes into its own not a moment too soon.

CMMS, PDA, data entry, Wireless	Hundreds of vendors have been working overtime to improve their systems. In the next generation, look for focus on the data entry side. Most of the money is spent on entry of data and it is the area of most problems. More interesting PDA applications will be available. The next

	frontier is the availability of effective and easy-to-use shop scheduling software. Also look for gains in the ability of CMMS to deal with photos, documents, and video data.
Root cause software (RCA), RCM software, PMO software	Structured programs such as these and others can be used manually, or with spreadsheet and word processor support. It is a great time saver when excellent software is guiding the way. Look for more and more programs in these areas, supported by low-cost and easy-to-use applications.
Document management systems	Information is the key to maintenance. Imagine having every drawing, bill of material, technical manual, digital photo, wiring diagram, and sketch needed, available on demand at a desk or even on your wireless device.
Predictive Maintenance	Predictive Maintenance has been a booming field for the last decade. Look for lowering of costs and increasing intelligence in the probes. This technology will become increasingly part of the standard tool kit in the same way that a VOM is carried by an electrician.
Condition based monitoring	There is an accelerating trend toward packaging intelligence with the machine. Machines already read internal sensors to determine their health. Sophisticated copy machines call for service when their internal state sensors detect wear or deterioration. Look for more intelligence in the machine. Other applications of condition-based maintenance will proliferate and begin to replace periodic inspections. Machines will report impending issues over the plant-wide data network and show up as work requests in the CMMS.
Laser Alignment	Tools to aid installation, such as Laser alignment gear, have reached high levels of intelligence and will continue to improve. They will also become easier and faster to use.
GPS and automated routing	GPS has revolutionized tracking and finding directions for cars, rail cars, ships, and trucks. Look for real-time data on critical shipments of large spares from GPS transponders. For fleet operators, optimized routing has already saved millions of gallons of fuel and wear and tear on vehicles.

Enabling Technology for Lean Maintenance

Lubrication	Lubricants have been improving for decades and there is no reason to think the pace will slow in this competitive field. Nano-technology will be used to disperse additives to make oil properties more suited to the application.
Application of lubrication	Automatic lubrication equipment is improving, and getting smarter and more robust.
	Acoustic-coupled grease guns: The operator is now able to listen to amplified "acoustic" noise made by bearings going bad through a headset, detecting problems associated with lack of, or too much, lubricant.
Coatings	Coatings are getting tougher, more corrosion-resistant and greener. Look for advances from nano-technology to take coatings to the next level.

CHAPTER

12

Lean Planning and Scheduling

Benjamin Franklin says: By failing to prepare, you are preparing to fail.

Planning and then scheduling maintenance work has demonstrated a capability for significant increases in productivity. That factor alone recommends it to the Lean Maintenance Hall of Fame. The reason why Planning and then scheduling is so effective is simple to say but difficult to accomplish.

The easy part is that scheduling the planned work allows the planner to work out the problems (generally with absent or conflicted resources.) ahead of time. Working out problems ahead of time saves 3 to 5 times the time invested, in reduced execution time. In terms of both forced and planned shutdowns, the savings could very well exceed 50 times.

The more difficult task is to manage the thousands of details that make up the resources of all the maintenance jobs running currently and in the near future. We have to manage the labor and skills, parts and materials, tools and equipment, access and permits, for the asset to be worked on, and finally to provide any required drawings, diagrams, or specific information.

In a modern maintenance department, this information is usually held in the CMMS database. Keeping up with this data is a Sisyphian task that is never done. The details available are always sketchy and imperfect. Even the scope of work is not always known until after the job starts. Yet planners have struggled against the odds to create effective plans that help the workers get the right stuff to the job.

Lean Planning and Scheduling

This particular Lean project is not one to be taken on lightly. In fact it could be argued that since it is a large change of business process it is not an appropriate Lean project at all. Planning and scheduling clearly facilitate Lean maintenance practices so we will discuss it here. Broad agreement from not only production and operations, but also from top management, is needed. It is a year-long program to get started, and it requires extensive investment in labor and training, and a working cooperation with the stockroom. Even with the elements in place, getting there takes some fortitude.

In addition to the important job of improving the amount of work generated by a given-sized crew, we have the additional power of being able to execute other Lean projects by planning and scheduling them. Because planning and scheduling increases production, there will start to be openings to insert some of the Lean projects and still get the work done to keep the plan operating. One of the best arguments for having a schedule is to provide resources for Lean projects without disrupting essential maintenance activities.

A study done in the late 1980s really tells the story of the source of the productivity gains from effective planning and scheduling.

Typical Maintenance Worker's Day—Reactive (without planning and scheduling) versus Pro-Active (with planning and scheduling)

Study done of 25 heavy-industry plants in the USA in 1986, supported by subsequent studies	Reactive without planning and scheduling	Pro-active with planning and scheduling
1. Receiving instructions	5%	3%
2. Obtaining tools and materials	12%	5%
3. Travel to and from job (both with and without tools and materials)	15%	10%
4. Coordination delays	8%	3%
5. Idle at job site	5%	2%
6. Late starts and early quits	5%	1%

Chapter 12

Study done of 25 heavy-industry plants in the USA in 1986, supported by subsequent studies	Reactive without planning and scheduling	Pro-active with planning and scheduling
7. Authorized breaks and relief	10%	10%
8. Excess personal time (extra breaks, phone calls, smoke breaks, slow return from lunch and breaks, etc)	5%	1%
9. Subtotal	65%	35%
10. Direct actual work accomplished	35%	65%

Each 1% of the above chart represents 4.8 minutes of the day. The 35% direct work in reactive mode versus the 65% direct work in pro-active mode provides a simple but clear justification for establishment of a Planning, Coordinating, and Scheduling operation (even in a Lean corporate environment).

	Detailed explanation of the items on the above chart **Typical Worker's Day**
1	Represents 24 minutes a day for no planning. Simply having someone look through the job, answer questions on paper, decide on parts/tools, and see if permits are needed, saves time in the communication between the worker and his/her supervisor. Having the package saves almost 10 minutes.
2, 3	In almost every maintenance department, significant time is lost in extra trips to and from the job, and to the tool crib and warehouse. Having the tool or parts list made out ahead of time can save a couple of trips, which translates into almost an hour saved a day.
4	The difference here is between asking the electrical foreman for an electrician tomorrow 'sometime' and asking for an electrician tomorrow at 9:30 am. Coordination delays include waiting for other crafts, lifting gear, permits, cleaning, custody, decontamination, and lock out.
5, 6, 8	From 5 on down the savings are from the human tendency to please bosses and from pride in getting the job done. What is observed is that people who are given a reasonable amount of work for a day will

	Detailed explanation of the items on the above chart Typical Worker's Day
	try to finish it. They leave early less often, are idle less, and take less extra time off during the day.
10	In practice 50% is more realistic but that is still a large jump in productivity. There are benefits below the waterline that are not listed here, but they make the case even more compelling. Planning and then scheduling those jobs Leans up the maintenance department by a greater percentage than almost any other activity.

Another way to appreciate the advantage of job planning is to depict what happens within an individual job without planning. Take the perspective from the point of view of the maintenance worker. Technicians frequently jump into the work without forethought. Shortly, they encounter a delay for lack of a spare part, tool, or something else that is missing. This sequence of working until they get stopped may be repeated several times before the job is completed (if it can be completed). In the planned mode, the needs are anticipated (as best as they can be) and are listed before a technician is assigned. The technician can just read the list and pick up what is needed before starting (and even more importantly, not even start if the resources are not available).

The comparison is graphically presented below. Remember, each dollar invested in planning typically saves three to five dollars during work execution, and the duration of a planned job is commonly only half as long as that of an unplanned job.

Professional Planning Versus Planning on the Run
Planning on the run

Professional planning

Note that the best technicians already do their own planning. All technicians who moon-light doing maintenance work as a contractor get pretty good at planning (or else they don't earn much and would do better by working overtime).

Plannable work comes from pro-active programs such as PM, PdM, condition-based maintenance, and TPM. The inspection parts of these programs detect impending failures before they occur. These impending failures are written up as corrective jobs, and are entered into the backlog. In such an environment, the bulk of the maintenance workload should be plannable (meaning there is enough time to plan the job). and most or all of that work would be planned. The next step is to choose the jobs to be scheduled with operations, and then finally, to schedule the planned jobs.

In a shop dedicated to Lean, the scheduled jobs make up the bulk of the weekly workload or 70 to 80 % of the available maintenance resources.

Steps in planning

> ➢ Determine the scope of the work. If the job is not familiar to the planner, or it is otherwise advisable (because of changes since the last trip, hazards, complexity, etc.), a visit to the job site and a discussion with the requestor might be important. Pinning down exactly is to be done is sometimes half the battle.
> ➢ Summarize the required budget for the job, thus defining *"how much"* it will cost, and obtain whatever authorizations will be required.
> ➢ A preliminary go-ahead is usually given at some point early in the process. If the job is a routine repair (even a large one) the level of approval might be casual. If the job is new, special, or outside the regular business, approval might be very formal, with top management signatures required.
> ➢ See if the job has been done before and if so, review the old job plan. If the job has been done before, adapt the job plan to the new circumstances and you might be done with planning right there.

Lean Planning and Scheduling

➤ Taking the physical location and spaces around the equipment into consideration, plan the manner in which the work is to be accomplished, thus defining *"how"* the job will be performed. List the steps to be performed. Include ideas about how many people can work on the job efficiently, and consider the movement of materials into and out of the area.

➤ Establish duration and manpower needs required to *"perform"* the work. Determine the skill sets needed, and any special licensing requirements (i.e., certified welder). Clarify the sequence of skills required throughout the job. Will the work be done in-house, will a contractor be called in, or will some combination of the two be needed to do the job? These steps define *"who"* is to do the "what."

➤ Next, identify all spare parts, materials, consumables, special tools, PPE (personal protective equipment), scaffolding, and equipment necessary to do the job. Determine if the items are in stock. If not, determine where and when they can be obtained. The latter step should be accomplished with the support of Purchasing.

➤ Identify essential reference materials and include them in (or link them to) the Planned Job Package.

➤ Finally, planning is not completed until everyone knows what is going to happen. Create a *Communication Plan* with all parties involved as well as sending to management, the plan of WHO is to do WHAT, HOW, by WHEN and for HOW MUCH.

A planning package may include items shown below. (larger, more-complex, high-hazard, and jobs you have less experience with, might have more support information):

Work order
Work planning sheet
Job plan with details by task with step-by-step procedures. Time for each step (task), summarized by resource group and for the total job. Labor deployment plan by craft and skill including labor-hour estimates (for very large jobs). Consider contract as well as in-house resources. Consider use of a GANTT

bar chart or CPM/PERT network chart to help plan task sequencing for assigned crews (important). Do everything possible to minimize the time the asset is out of service.
Job Hazard Safety Analysis looks at all hazards of the job and seeks to remove or mitigate the hazards.
Pre-shutdown work list (use of prefabricated parts)
Bill of Material. List all materials needed for the job, including an acquisition plan for major items. Determine if the material is authorized inventory or a direct purchase item. The planning package should include spares, reservations, and staged locations.
Requisitions
Materials control list
Shutdown protocol and start-up protocol
Site set-down plan (where to put everything used for major tear downs)
Lift plan
Scaffolding
Copies of all required permits, clearances, blind lists, and tag outs
Prints, sketches, digital photographs, special procedures, specifications, sizes, tolerances, and other references that the assigned crew is likely to need

Development of Work Programs

Work Programs are the vehicles whereby maintenance resources are perpetually balanced with the maintenance workload. Without this balance, scheduling is impossible. The Work Program determines exactly how many hours will be made available for planned backlog relief. To prepare the program, start with a series of deductions and calculations:

❑ Gross labor hours authorized each week. This value is simply the number of hours per person to be worked next week times the number of workers.

❑ Deduct hours not available for maintenance work including vacations, absenteeism (use historical percentages), training, meetings, etc. This sub-total is subtracted from Gross hours.

❑ Deduct the average hours spent responding to urgent conditions such as equipment failures that cause production downtime, or urgent jobs that management wants done immediately.

❑ Make a list of all PM and PdM scheduled for the next few weeks. Calculate the number of hours that would be needed if all pro-active work was done on time next week. Then deduct all projected PPM work for next week.

❑ Add in any hours budgeted for overtime and any authorized contract support.

❑ Identify the resources available for backlog relief. This is the number that the planner carries into the coordination meeting.

❑ One important calculation is the number of weeks of backlog. The available hours for next week are divided into total hours of "Ready" and "Total" backlog to quantify backlog weeks compared with established benchmarks.

Scheduling

The first step of scheduling after the work program is complete is coordination. The planner will first run the Ready backlog (all jobs that have been identified that are ready to go). The ready backlog is then validated (cleaned up). All materials, engineering, rental equipment, and authorizations for these jobs are in place. The ready backlog and the scheduled PMs for the next schedule period are presented to operations. The work program that represents the most accurate number of hours available for work is also presented.

Coordination is really a meeting, the outcome of which is a work list for next week. The scheduler creates the schedule from the work list. Before the actual coordination action items come up in the agenda however, other things must be dealt with.

Agenda for the Weekly Coordination Meeting

❑ Item one might be "How did we do against last week's schedule?" and exploration of reasons for non-compliance. What jobs were not completed as scheduled, and why? Did maintenance fail to get to them? Did Operations deny equipment access? Did unscheduled jobs break into the schedule, and were they necessary? Such questions highlight underlying reasons for poor schedule compliance. When it is the Operations Manager who raises the questions, the process is more effective, problem resolution is speedier, and future schedule compliance improves rapidly. The Manager should therefore attend the coordination meeting at least once a month.

❑ The beginning of next week's schedule, showing resources available and demands already established (PMs, Corrective Work, Carry-overs, Project commitments, etc.) should be listed.

❑ Next, the Production Schedule should be presented to clarify any operational support required from maintenance (set ups, change-over's, change-outs of production expendables, etc.) as well as equipment access windows that can be utilized by Maintenance.

❑ Critical jobs from last week (or before), that are delayed for lack of parts, engineering, approval, budget, or other reasons, and considered for possible expediting should be reviewed.

❑ At this point, all parties are ready for the give and take as to whose jobs are next most important in the optimum interests of the overall operation. This discussion continues until all available resources are committed in general terms. Some changes are inevitable when the scheduling process addresses the specifics.

❑ If participants continue to lobby for more support, overtime and contractor support is the next consideration. These decisions will require approval and funding.

Scheduling

The scope of Scheduling includes:

❑ Bringing together in precise timing the six elements of a successful maintenance job: labor; tools, materials, parts and supplies; information, engineering data and reference drawings;

custody of the unit being serviced; and the authorizations, permits, and statutory permissions.

- ❏ Matching next week's demand for service with the resources available after accounting for all categories of leave, training, standing meetings, and indirect commitments, plus consideration of individual skills.
- ❏ Preparation of a "Weekly Schedule" that represents the agreed-upon expectations regarding planned work orders to be accomplished with the available resources. The schedule also assures that all preventive and predictive routines will be accomplished within established time limits.
- ❏ Consideration of alternative assignment strategies where the schedule assigns specific jobs to specific people (allowing second-string players into the game to gain experience … as feasible.

Part of the planner/scheduler's challenge is ensuring that responsible supervisors receive and understand the planned job packages for scheduled jobs.

13

Lean Shutdowns, Outages and Turnarounds

There is a special case of large maintenance jobs where the plant must be shut down, and that is the planning and scheduling of shutdowns and outages. The Lean opportunities are enormous. These massive events offer two opportunities to Lean zealots. On one side the shutdown itself is the largest event in the maintenance year. In a major cement producer 40–60% of the maintenance budget for each of their over-100 cement plants is spent on shutdowns.

By its very nature, the shutdown is fat. Small improvements in managing shutdowns can provide large weight losses for the maintenance department. Shutdowns are fat because the attitude is get 'er done and we'll worry about the budget when it's over. This attitude is good because downtime is so expensive, but sometimes it is taken too far and in itself becomes a fat approach.

Some recommendations:
➢ Keep an eye on over-ordering of materials and return unused stuff as soon as you are clear you won't use it.
➢ Note explicitly whether there are enough supplies for the whole shutdown. Supplies include rags, oil-dry compound, welding rod or wire, gases, nuts and bolts, etc.
➢ Keep an eye on excessive numbers of rented cranes, welding units, tanks, scaffolding, and other equipment. Investigate and return what is clearly not needed and doesn't provide any benefit, unless it is there to provide insurance against some loss.
➢ Return rentals of all kinds as soon as practical.

Lean Shutdowns, Outages and Turnarounds

- ➤ Be on the look out for situations where resources are being paid for but because of resource leveling problems they are not being used. Have some lean projects to use them.
- ➤ Look at the scope of work as a contractor would. Be sure it's as good as possible. A better scope will result in lower prices if there are fewer unknowns.
- ➤ Settle claims with your contractors promptly.

On the other side is the opportunity for specific improvements that are engineered for and built during the shutdown. Shutdowns are the single greatest opportunity for lean initiatives.

First, it is important to plan each job and schedule all the jobs as completely as possible. In larger events it is advantageous to use project management software such as Primavera, SAP PS, or Microsoft Project. If we go back to the beginning of the computer age we will find shutdowns were one of the first events that took advantage of early project-management software.

In the late 1950s there was a race (although most of the work was secret, so there is some question if any of the teams even knew the other was working on the same problem), for the first computerized project-management application. IBM, Booze, Allen and Hamilton, and Lockheed are generally given the prize for their development of the PERT (Project Review and Evaluation Technique) computerized project management system for the Polaris Missile program, and the building of the nuclear submarine USS George Washington.

But at almost the same time, DuPont and Univac (now Unisys) came out with the CPM (Critical Path Method) system. Eventually CPM became more widespread because it gave most of the advantages of PERT with a fraction of the work. The first shutdown using project-management software was of a DuPont Catalyst Plant in Kentucky in 1959. The duration of the shutdown was reduced from 125 hours to 97 hours. This result was a huge return on investment, and one of the first examples of computers used as an enabler for Lean Maintenance.

So the first lesson is that almost any manually-scheduled shutdown (over a certain size) can be optimized by proper use of project-management software. The size might be more than 25 to 50 activities, and sometimes smaller. Of course PMS won't make a fat shutdown Lean, but it will help.

Shutdowns almost always are relatively wasteful. Almost always because of some of the facts about the way shutdowns are run. If you have a high-downtime project it pays to have resources available, even if you don't use them. If it could shorten the shutdown by three hours, the fact that you have all these resources there, a lot of times doesn't make any difference. You know that, as far as the budget goes, it's intrinsically a costly event. As mentioned, in some plants over half of the maintenance budget is consumed by shutdowns. The overtime, the amount of management headache, in all ways makes a shutdown a costly event and generally a wasteful one. In one instance it was found that any jobs that could be completed outside the shutdown period were less than 1/2 as expensive for the same scope of work as during the shutdown.

One of the most costly components is discoverables. These are jobs found after the shutdown begins. Looking at this aspect from the point of view of waste, what can be done to manage discoverables?

Discoverables go on a diet

> Open everything on day one.
> Keep a history based on previous experience. Such a history will be important because it will show the deterioration in efficiency, if any. A lot of things deteriorate at a relatively constant pace because of what's going through all of them. And they have similar failure modes.
> Diagnostic technology might give an indication of what's going on. Schedule NDT right before you close the work list.
> Pre-shutdown. Maintenance workers on an oil platform out in the Gulf of Thailand do a mini-shutdown before they do their big shutdown. When they do they mini-shutdown they open everything up; inspect it, close it up, and go back into service. Now, that's not possible in a lot of places, but in this situation the crew said that their shutdowns were relatively controlled and didn't produce a lot of surprises. The amount of discoverables also was dramatically lower, but the crew still had to

convince management. Could you imagine this, "Well, we want to do two shutdowns instead of one? You know, in order to manage the one, we're going to do another one."

➢ Another Asian group did a dry run. A pharmaceutical company hired a contractor to do a complete dry run before the main shutdown. They acted everything out before they touched the equipment, literally bolting stuff together, and soldering it to get it into the room. The company paid for it but, when that shutdown started, it was like a choreographed dance troupe.

➢ The same company hired the shutdown contractor to draw up and start to plan the next shutdown while they were doing the dry run. With all the equipment opened up, the contractor could make measurements, do sampling and testing, and even count gear teeth to get a jump on the next shutdown.

People get trained in the project management body of knowledge, which is one aspect of running shutdowns. It turns out that's what is needed to learn to run the critical job, the one job. But there's a hundred other jobs running that have to be managed in some other way. These other work orders are confusing and may make the problem more complicated because there's no logical connection to the 'big' project. In the language of project management, the unrelated work orders have dangling beginning and end points (logically unconnected from the rest of the network diagram used to manage the shutdown).

One issue that promotes fat is management's inability to set a schedule and stick to it. The start date is picked and everything is put into motion for that date. Suddenly it is found that the shutdown is two months earlier than was expected.

Remember, what we're trying to do here, is to identify lists of things that involve waste. Shutdowns are a huge area of waste. But a whole lifetime of study is needed to get good at that. People who do shutdowns know that this learning is non-trivial, but getting a little better at it can shed much fat and is a good thing.

Projects:

Projects are another area of waste. A lot of waste goes in with any new project that looks like saving money. The plan makes it look as if money is being saved but actually, waste is being created. Waste is in a few big areas:

Parts: The project orders all kinds of materials and other resources and they are left over at the end. The stuff should be returned if possible. If it can't be returned, what do you do with it? To cut the costs of some projects, the leftovers are put into stock and credits are issued. The problem is that the parts may never be needed again. They are just absorbed into the inventory and then got rid of, years later.

Value engineering: Every project seems to reach a point where management wants to squeeze some money from the budget. That saving is not a problem, and could be considered a Lean activity in itself. It becomes a problem when the items that are value-engineered out will cause maintenance problems for years to come. The value-engineering process should look at the life cycle costs of its decisions.

A notable example of value engineering was in a 14-storey hospital tower. When it was built, the project ran over budget. To save money it was decided to eliminate the water shut-offs to each floor. Now, whenever there is a problem between the fixture shut-off and the main shut-off (like adding a water outlet or servicing the line) the water in the whole building has to be shut down. On one visit they were experimenting with a line freezing unit (making an ice plug so that the line could be drained without shutting off the water supply to the whole building).

Half baked projects: One of the fattest project mistakes is to go in with inadequate planning, money, time, or expertise. The project team then flees the scene with the job 90% done and the maintenance department is tasked with making it work.

Meetings during shutdowns and at other times

A lot of companies have very bad meeting habits. People come in late, or don't do their homework. They do not pay attention and then act inconsistently with the decisions of the group. They don't have good discipline around meetings, and management often doesn't model good meeting behavior either.

Lean Shutdowns, Outages and Turnarounds

Wasted meeting time is highly leveraged. If there are 8 people at a meeting and they are waiting for the ninth, then the time of all eight people is wasted. It isn't just the one person's time—the loss is leveraged so that the time of eight people is gone. However, meetings are essential and are the bread and butter of the shutdown. You've got to have meetings. The Lean project here might be to train people in better meeting practices to make the meetings more effective.

CHAPTER

14

Lean Fire-Fighting

Benjamin Franklin says: The worst wheel of the cart makes
the most noise.

For the purists, Lean Maintenance means there are few if any
breakdowns. Most of the causes of the breakdowns have been dealt
with in the PM system (wear out) operator training, or in machine
operation simplification (Poke-a-yoke or mistake-proofing). In the
real world, machines still break. Can we adopt an attitude of
Leanness toward breakdowns?

Is it possible to make fire-fighting lean (or at least leaner?). The
answer is yes. To figure it out, do this thought experiment: think
about whom in our society is the best at dealing with emergencies?
A few years ago we visited a local fire station with the kids. The
issue with fires is to get to them as soon as possible with the right
tools and supplies. Fire fighters have worked and re-worked this
problem over the years.

On the floor outside each of the doors of the fire trucks was a set
of boots and a (specific) pair of fire fighter's overalls, set up so that
a fireman could step into the boots and pull up the overalls. The
jackets, helmets, air packs, and entry tools, were on hooks in the
truck. The truck's water and diesel tanks are always full, and the
trucks are always ready for action. Can you imagine the cost in lives
and property damage if each fire fighter had to stand in line at the
parts window for his/her helmet or air pack? This fire station
approach points to good ideas we can use in our factories, to make
our emergency service Lean.

Lean Fire-Fighting

The fire truck has compartments for most of the tools that are used in the typical fires known in the local area. There is not much casting around for a tool for entry or climbing. The firemen have also developed standardized multi-purpose tools over the years.

As with fires, emergencies in your plant can be planned for ahead of time, but they cannot be scheduled. Many emergencies have either happened before or can be reasonably predicted, based on the plant equipment, raw materials or processes. Planning in this sense means that the resources (parts, tools, supplies, equipment, etc.), needed for the repair, have been identified ahead of time. The job plan is stored in the planning library, or in the CMMS.

For example, if the plant has pumps we can reasonably assume that a catastrophic emergency pump failure will eventually occur. We can have job plans for replacement or rebuilding of all the pumps in the plant in the planning library (completed ahead of any breakdowns). We can also do a criticality assessment (part of the stocking policy determination) to determine which parts or which entire assemblies we should carry in stock to support emergency repair.

Of course, in the fire house, they don't know when they will get called for a fire but they do know that they will get called. They also know something about the qualities of most of the fires they face (given the buildings and businesses in the sector they service). The firemen have rehearsed scenarios for most situations, and they have a very well understood and well known chain of command. They also have standardized and multi-use tools and equipment, packed and ready to deploy.

Plant emergencies need a clear chain of command, just like a fire scene. There are certain tools that might not be justifiable for every-day use but would be life-savers for a breakdown. We could design and deploy kits for emergencies in different parts of the factory.

Another great organization that deals directly with breakdowns is a typical hospital. Have you ever been to the emergency room in a hospital? The first step in an emergency room is triage, which separates the heart attack patients from those suffering from gastritis, brought on by overeating.

Chapter 14

There are tool boxes that are dedicated to certain breakdown scenarios. When someone comes in with a broken bone, the medical staff rolls over the orthopedic cart with all the tools and materials for broken bones. They cannot fix everything with what is on the cart, but they can stabilize almost everything and fix upward of 90% of the emergencies.

ERs have whole carts or drawers containing specialized equipment for different kinds of emergencies. No-one fumbles around, looking for the critical drugs, when someone's heart stops. Doctors and nurses do not stand in line at the pharmacy for an EPIPen for someone having an allergic reaction, and who can't breathe (but they must order and wait for medicines not used as often such as snake venom antidote).

One practice that hospitals have down to a science is replenishing the carts after an emergency. It is a nurse's or aide's job to look through the drawers and count the materials that have been used, and replenish all the consumables and drugs (for billing purposes too). They also make sure that any tools that have been used are sanitized or replaced.

Many factories go to the trouble of having carts or job boxes for emergencies. Unfortunately, parts that are used during a breakdown often are not replenished, broken tools are not mended, and supplies are borrowed for minor jobs and not reported or replaced. It is usually no one's explicit job to replenish the cart, change the batteries, and clean up the mess in the drawers.

Unlike hospitals that only have to deal with one make (human), and 2 or 4 basic models (female, male, and kids of both genders) of different ages, most factories have a very wide variety of equipment, so it would be wasteful of space and parts to have carts for everything. What would be the Lean response?

Lean maintenance calls for the least-intensive response possible, with nothing extra and nothing missing. It calls for cross-trained crews (so that the first responder is likely to have the skills to fix the problem properly). It calls for parts to be stocked for the most critical, the most dangerous, and the most common emergencies nearby.

Lean Fire-Fighting

Every area of our lives, where quick response is essential, follows some basic best practices:

1. Have the tools and materials needed for 90–95% of the possible incidents in a cart, truck, box, or other segregated area. The tool box should be mobile and easy to move to the breakdown. Carts are okay in a building, but in a large industrial site consider footlockers, tool boxes that can be thrown onto the back of a truck, or even complete trucks. An alternative is fixed cabinets containing emergency spares, and (if it is justified) with specialized tools. The cabinet should be locked, but the key location must be well-known and accessible whenever the plant is in production or maintenance is at work.

2. The tools and materials are put away into the same places, pockets, drawers, and cabinets each time they are used. In a breakdown it is essential that everyone knows where everything is. That is why the drawers of all a hospital's crash carts (used when someone stops breathing, or their heart stops) always contain the same tools or drugs. This rule is so important in a hospital that there are periodic task forces that review and redesign the standard layouts as better technologies and tools are introduced and become more commonplace.

3. Care is taken to clean, lubricate, charge batteries, and generally care for the tools after the crisis is over. There is nothing more frustrating than being in the middle of the next crisis repair and having a dead battery on a screw gun, meter, etc. Parts and materials used must be replenished as soon as the crisis is over. As mentioned, in a hospital, a nurse is assigned to take inventory and replenish the cart after each use. Who is explicitly assigned to this task in your facility?

4. Create a work order listing what was done, what was used, for the records. Part of every emergency job is to clean and put away the tools, replenish the used parts, and fill out the paperwork.

5. On a large, dangerous, or expensive breakdown, consider having a meeting a day or two later to discuss what happened,

what went well, and what needs to be improved for next time. This meeting is not to point fingers but rather to identify what worked and what didn't. Hospitals have a committee called the M&M committee (Mortality and Morbidity) that reviews every death and sickness (that started after admittance to the hospital), for what can be learned, and to determine whether any procedures need to be changed (or reinforced).

6. Spend time training people in how to respond to breakdowns. Set up a coherent chain of command and be sure that everyone knows their role. Let the people that are great at fire fighting teach what they do, and how they approach these types of repairs. Use the work orders from the last crisis to jog people's memories.

Cart design: The fire fighter's cart should be intensively studied by both the maintenance personnel and management. Consider the procedures of the local phone, gas, water, cable, or electric companies. Tremendous thought goes into how to outfit a service person's truck. Next time you have an opportunity, ask the telephone installer or gas repair person how their truck is set up and why. Apply the lessons to the fire-fighting cart, toolbox, jobox, van (or even 5 gallon bucket!). The more often you have the needed part in the cart, the more downtime you avoid and money you save (and the Leaner you are).

Organizations that are serious about quick response to breakdown do the following:

1. Have meetings on this topic and discuss what happened in the past with their old-timers.
2. Include the maintenance customers in these meetings.
3. Decide who will do what, when an asset breaks down.
4. Decide what actions the operators can and should take.
5. Decide where you will keep the cart and the spares.

15

Lean PM

Benjamin Franklin says: A stitch in time saves nine,

The Maintenance Department's contribution to the success of the enterprise

Within organizations there are experts in many disciplines. As the organization gets bigger, this aspect becomes both more important and more economically smart. Expertise is important because small mistakes can be enormously expensive, particularly in a large organization. As an organization grows, the probability of issues arising becomes larger because the population of events is also larger.

For example, having an expert in managing cash to squeeze the most interest out of deposits and reduce incoming float (and increase outgoing float), is not very important in a company with a $1,000,000. The difference of even 2% in interest over that amount would amount to less than $55 a day (which would not pay for the effort). But for a large enterprise with a billion dollars, the same advantage of having an expert get 2% more could be $55,000 a day (which will pay for the effort). The same considerations apply to risk management, energy conservation, safety, and a myriad of other areas.

Every one of these experts makes a contribution to the success of the enterprise by lending their expertise to the problems and opportunities of that company. Legal experts strive to contribute to the company by protecting it on the legal front. Risk managers contribute

Chapter 15

by eliminating the probability of catastrophes, and adequately insuring the company in case the worst does happen.

Companies occasionally ignore their own experts. Sometimes the company lawyer says a deal should be structured this way or that way, and the executives decide to ignore the advice and structure the deal their own way. The executive management has the right to ignore the warnings of the experts and go their own way, for good business reasons, of course.

The same idea applies to maintenance. Maintenance managers, supervisors, engineers, and trades people are the companies' experts in reliability and machinery health. If you want to find out how to operate to avoid problems, increase the length of time between failures, or establish routines to care for your assets, you look to the maintenance department. In a lean operating environment, the authority on machine health is the maintenance department. Reliability knowledge is stored in people's memories, and in the company's PM system. Lean operations should always look to the PM program for ideas on Lean projects. PM can be Lean if it is handled correctly.

PM is sometimes fat

But PM systems are sometimes notoriously fat! The first consideration is that in PM, the most common fat is too-frequent PM. Some PMs are done for years without showing any benefit. Operating conditions may change, but the PM tasks do not. Lubricant life may improve, but the drain intervals are not changed.

The second fault is ineffective task choices. Doing PM tasks that are not related to failure modes is Fat. Doing tasks too infrequently, so that the tasks cannot catch the breakdown, is also Fat. The tasks should be designed to be performed at the proper depth, frequency, and level of technology.

The third item is a constellation of people problems. This fault includes lack of training where the PM person wants to do the tasks but doesn't know enough to do them correctly. Another drawback is the lack of desire to do the work as designed (workers do what they think should be done, regardless of what is written on the task list). Of course a small percentage won't do the PMs at all, either due to a poor attitude or to feeling that PM is demeaning. Another small percentage does not have the strength, visual acuity, or intelligence, to perform the task as designed.

Don't fix it if it ain't broke

In some plants PM has a bad reputation. Operations and production shield themselves from the PM crew. The PM crew is shunned because they cannot reliably return the asset to service after PM activity without problems. All that said, when correctly applied, PM is a great area to look at for Lean Maintenance opportunities.

Maintenance is no different than any area involving expert advice. Sometimes the experts are wrong or ignorant. The Lawyer, or Accountant, or other expert, is incompetent, wrong, or just lazy. If the truth is told, some maintenance "professionals" are also incompetent, wrong, or just plain lazy.

Lean action item 1: Count the number of Corrective items you find when you do the PM tasks. A Lean PM frequency is 5 to 6 PMs for each corrective task.

Lean action item 2: Automated lubrication equipment can be extremely Lean. The new systems can be put into place for long periods, and will deliver lubricant reliably without extra labor. These systems are particularly useful where the grease points are hard to get to, in dangerous or distant locations. Many systems that feed multiple points will set a flag or an alarm when the points get too easy (cut line), or too hard (plugged up), to pump to. Of course, automated lubrication equipment has to be added to a PM list because the lubricant does get used up and the system is not infallible.

Why consider Automated Lubrication Equipment?

- The equipment can be set to deliver the right amount, of the right kind, at the right time.
- All levels of technology from simple to complex, are available.
- There are many pricing levels.

Chapter 15

- Once installed (properly), risk of contamination is reduced.
- Places that cannot be reached at all by hand can be lubricated.
- Places that cannot be reached safely by hand can be lubricated (without interrupting the operation of the machine).

Even if PM is fat, organizations love PM

The way that the maintenance department expresses its expertise is through the PM system. The PM system is the combination of the expertise of the original equipment manufacturer, the skilled trades, the maintenance engineer, and the supervisor. This expertise is directed toward the most expensive, the most dangerous, or the most common failures.

Everybody loves PM. Doing PM shows the world that you care about the assets. In some areas, PM is treated like a religion and never questioned. People believe and have faith in PM. But is PM Lean? If it is fat, how do we make it lean?

To make the question more complicated, maintenance people have different and often contradictory definitions of PM. In most places, PM is a series of tasks written on a task list, which either extend the life of an asset or detect when an asset has suffered deterioration and is on the road to failure. Although the PM system provides for finding any deterioration, the PM task list does not include such activity. Fixing deterioration found by PM inspection is called Corrective Maintenance.

The first aspect of PM is life extension. Overwhelmingly, breakdowns are caused by a lack of (or too much) lubrication, excessive contamination, and loosening or missing fasteners. Some studies show as many as 3 out of 4 breakdowns come from these causes. It is pretty clear that time invested in basic maintenance reduces breakdowns.

The question is, does it pay (in a rigorous financial sense) to perform basic maintenance activity? Here we can borrow some of the ideas of RCM (Reliability Centered Maintenance described in more detail in a subsequent chapter).

RCM has its roots in the aircraft industry after World War II. There was an urgent need to increase the reliability and reduce the maintenance costs, if air travel was to become universal, safe, and profitable. RCM is a rigorous, structured, process, designed to identify component functions, losses of function, and failure modes. Those requirements have led to breakthrough thinking for aircraft, and for the other areas where it is appropriate.

The part of RCM relevant to this discussion is its orientation toward consequences. In RCM, breakdowns are not the major issue. The major issue is, what will be the consequences of the breakdown or other event. Consequences are divided into 4 types:

- Safety
- Environmental
- Operational
- Parts and labor

If the consequence of failure is one of the first two (Safety and Environmental), the designer is compelled to redesign the asset to eliminate that mode. The other choice is to design a PM/PdM task that will either eliminate the probability of the failure happening, or require a task rendering that failure mode detectable well in advance of failure. If the consequences of failure are of the operational, or parts and labor (the last two consequences) type, then the whole discussion is economic. The question to be answered is, how much does the breakdown cost versus how much does the PM task cost over the same period.

PM activity in the domain of safety and environmental is, for the most part, exempt from the Lean discussion (in other words don't mess with safety and environmental inspections). It is difficult to put a price on an industrial accident or death. We would not want to be in any discussion about saving 100 hours a year by skipping the overhead crane PM and lubrication, given that the probability of failure and injury without the inspection is 4 or even 10 times higher.

Is basic maintenance Lean or is basic maintenance worth its cost? In a production environment with expensive downtime, generally, basic maintenance is worth it. Either the basic maintenance itself, or actions that automate basic maintenance (like automatic lubrication), are worth while. Basic maintenance has other advantages beyond immediate economics because the group that does the tasks gets intimate knowledge of the equipment. That knowledge can be leveraged into higher uptime and used in the resolution of minor problems.

The second part of PM is inspection to detect deterioration before it reduces the function of the asset, and certainly before it causes a catastrophic failure. Inspection comes in two flavors: human senses, or low tech, and using instruments, or high tech. High tech inspection with instruments is by convention called predictive maintenance. We will deal with Predictive Maintenance in another section.

Preventive maintenance inspection can be evaluated the same way as basic maintenance. If the inspection is designed to uncover a failure mode that will cause safety or environmental catastrophes, it is beyond the scope of this discussion and should not be included in the Lean analysis. But if the failure mode will cause downtime or loss of parts and labor, we would subject that task to economic analysis. The question is the same. Is the task worth it? Note that the analysis is properly done on a task-by-task basis. It also should be done for the whole PM task list to determine if PM is the right strategy at all.

Fat to Thin

There are techniques of reviewing PM task lists with an eye toward taking out some of the fat. The most notable technique is PMO (PM Optimization). PMO could be called RCM lite (with apologies to all involved). The designer, Steve Turner, a long time RCM expert, has taken the essential elements out of RCM, used them to analyze PM efforts, and packaged them into a usable and understandable process.

The process involves collection of the task lists (some lists are not even written) from everyone who touches the equipment, including operators, maintenance workers, calibration people, tool changers, even housekeeping. These lists are compiled into one master list. Each line of the task list should point to some failure mode. The task should eliminate the failure mode, reduce the probability of that failure from happening to near zero, or offer detection of the impending failure in time to intervene.

A list is then drawn up of all the ways the asset can fail. The list would include both experienced failure modes and failure modes that are probable. Discussions with other users of the same equipment are also useful.

The two lists are compared. Each failure mode should have a task pointing to it. Each task should be pointing at a failure mode. When the process is complete, duplicate tasks have been eliminated, all tasks point at failure modes, and all failure modes are covered by tasks. The final step is to assign the tasks back to the original groups, based logically on location, frequency, and competence.

The result is a Lean PM Task list. The goal of Lean in the PM area is to make sure that the tasks have as little wasted time and materials as possible, and to get the biggest bang for the PM buck. Many PM systems have low hanging fruit (easy projects that will give a fast return on investment).

Condition-based maintenance

There is a field allied with Preventive and Predictive maintenance called condition-based maintenance or CBM. CBM provides an excellent opportunity to apply Lean ideas. In short, CBM instructs the maintenance worker to perform some series of tasks or activities only when some finite physical condition or a reading on a gauge or meter reaches a certain value.

The most common example of CBM is to change a filter when the differential pressure (difference between two pressure gauges) reaches a preset point such as 5 PSI. A pressure gauge is mounted in-line before the filter, and another is mounted after the filter. The filter change is done after an inspection, only on one condition. The inspection is done either by reading the two gauges by hand (or rather by eye) or by having a computer system read them. In either event, one pressure is subtracted from the other and the difference is compared to a preset value.

Leanness is provided by the fact that the filter is changed exactly when necessary. A traditional PM approach would have the filter changed periodically on a schedule, irrespective of its need to be changed. This approach is Leanest when the readings of the conditions are already being collected by the process control system. Then it is double Lean!

Chapter 15

PM is like advertising. You know that half of it is wasted effort. The great challenge is knowing which half should be eliminated? There is a pointer within your grasp that will point only at PM tasks that will make a difference. This pointer is the parts that you use.

Process: Review the parts that you use and make a listing of all the parts that failed because of breakdown. Sort these parts so that the most frequent failures lead the list. Divide those failures into three categories (this process will look somewhat familiar if you are a RCM devotee). The three categories are:

1. Parts failures where there is a safety or environmental exposure (failure could lead to death or serious injury or could lead to a serious environmental incident).
2. Parts failures where large costs were incurred in parts, labor, scrap, or downtime.
3. Most frequent failures not in the above categories.

Look at the parts and the failures. For all the parts in category 1, is there a task that would have eliminated the failure? After category 1 is complete, do the same analysis on categories 2 and 3. Always include the cost of the task in the review, and choose tasks that give a big bang for the buck! This type of approach can be very effective in tightening up a PM program.

Lean Idea for action: Add a series of tasks to the PM schedule on your computer system that initiate review of the 5 most-used breakdown parts. Have the analysis performed by a knowledgeable tradesperson, and give it a labor standard of 4 or more hours. This idea will be reviewed in greater depth in the section on finding waste.

16

TPM and Lean Maintenance

Benjamin Franklin says: Beware of little expenses. A small leak can sink a great ship.

TPM is a shift in attitude for almost all plants. On the surface (and the item most advertised) it is claimed that TPM uses the operators, in autonomous groups to perform all the routine maintenance including cleaning, bolting, routine adjustments, lubrication, taking readings, start-up/shut-down and other periodic activities. The maintenance department people become specialists in major maintenance, major problems, and problems that span several work areas, and trainers and operations become the specialists in machinery health.

Most TPM tasks are also PM tasks when they are being done by maintenance personnel. There's a lot of PM that can be done by operations, if they know what they are looking for and how to report what they see. The Leanness comes from having the right person do the right tasks, and the reduction of non-productive time (such as travel). Moving the maintenance effort to the section of operations that is located on or near the asset will be Lean.

For many of these activities connected with training, operations is the right resource for the job. The advantages from TPM flow from higher productivity, and from the effects of the shifts in attitude. The higher productivity comes from having the operator do basic maintenance. The operators are already at the asset, already have custody, and already have a job assignment. No additional travel is needed for the operator. The improvements in attitude come from the shift from a passive stance as an operator toward an active stance, becoming the person fully accountable for the whole thing, and an expert in machine health. Both topics will be covered in more depth in subsequent sections.

What is the balance between TLC and Lean? We have to look at the Return on Investment to determine where the balance is located. Too much TPM is the same as too much PM. It is Fat.

TPM naturally attacks the sources of ineffectiveness. Most companies spend enormous amounts of money on improvements in efficiency. Efficiency is defined as doing things the right way or getting the most output for the least input. TPM can be said to take the next step. TPM looks at doing the right things right. By attacking all the losses that impact production, TPM ensures that, at the end of the day, the pile of good salable parts made by the process is bigger. Sometimes, after a TPM implementation, the pile of good parts is a lot bigger.

TPM is an important part of traditional Lean Manufacturing. One of the few books in the field is called *Lean TPM, A Blueprint for Change—Harnessing Lean Thinking and Total Productive Maintenance* (by McCarthy and Rich). In the Toyota Production System (TPS) TPM and Lean are also extensively linked.

As mentioned elsewhere, the original orthodox Lean concepts were developed in Toyota's automobile assembly environment. Automobile assembly is a particular type of manufacturing. The manufacturing equipment such as assembly tools, robots, and small gauges, tends to be physically small. Many of the problems in assembly lines can be addressed by using a TPM concept, which is simply returning the asset to "like new" condition.

When you review the work orders in an assembly operation you see that most of the maintenance performed needs moderate skill on the relatively-small equipment (or extremely high skill that is beyond the scope of TPM). The range of skills is neither wide nor deep. Also, for the most part, the bulk of the work is safe and does not expose the worker to extreme hazards.

Given the environment of an assembly line it is completely logical that, in the Toyota Production System, TPM would be prominent. In fact, I would go out on a limb and say TPM would never have developed in a refinery, mine, or smelter, even though it is useful in those environments.

TPM can certainly be one Lean approach. It is one of the tools in the toolbox of Lean. TPM is not necessary to pursue the myriad of other Lean projects. Much of the waste in a typical maintenance operation has nothing to do with equipment effectiveness and will not necessarily be uncovered by TPM approaches.

The issue of TPM is that all the practitioners talk about TPM being a long-term project with implementation cycles that last years. TPM requires a radical change in culture that starts at the top and permeates all levels. TPM is executed on the shop floor. Don't get me wrong, this cultural shift is great for Lean maintenance. But, while TPM is going on in the background (with little return on investment for a year or more), the organization can be saving millions in the foreground with short, high-return, and hard-hitting Lean projects. In fact, Lean projects can be a practical motivator to stimulate the TPM culture shift.

Let us now look at some of the aspects of TPM and how they contribute to Lean projects, and examine how the organizational structure of TPM can support a long-term Lean initiative. The key to the long-term survival of either a TPM program or a Lean Maintenance program lies in the business structures such as reporting, and the incentives built around those programs to keep them going.

TPM has two missions

The first mission is described as a rigorous approach to achieve high machine utilization. The second mission, which is the more important one for Lean Maintenance, is a shop floor philosophy that is based on encouragement of operators to take a greater role in the health and productivity of the machines they are tending.

TPM has four elements

TPM is one of the most effective methods of improving the delivery of maintenance services while increasing the efficiency of the equipment. Although in its entirety, TPM doesn't apply in many situations, some aspects apply to all maintenance situations. TPM is the operations' department's extension to the empowerment, job enrichment, and total quality programs, on the production floor. The great advantage of TPM is that it can be incorporated into and can greatly enhance these programs. To begin with, TPM has the dual goal of zero defects and zero breakdowns. To achieve this goal, TPM has four elements:

1. The organizing principle is to maximize overall equipment effectiveness. Simply put, idle equipment is fat and equipment producing quality output is lean. TPM has a very strict definition of effectiveness. On a relative scale, every time you increase quality output from a given asset base you (by definition) lower the cost of maintenance per unit of output. Improving equipment effectiveness does not in itself make for leaner maintenance, but it certainly makes a Leaner enterprise. This relentless pursuit of increased effectiveness is a good place to stand for Lean Maintenance.

2. PM is both the source of excessive fat and of leanness in maintenance. As mentioned in the previous chapter, too much, too many, wrongly-focused, or poorly executed PM tasks are Fat and are time killers. TPM teams examine PM tasks. TPM shares the tasks between maintenance and operations (so that sharing in itself can sometimes be lean). The shared system of PM covers the equipment's complete life by taking into account both the life cycle of the equipment and the skill of the inspector.

To be Lean, PM should be modifiable, based on the life stage of the equipment. Without this modification ability, PM tasks might not reflect the failure modes of equipment at that stage and in that condition.

The divvying up of tasks and the accompanying analysis of task effectiveness can be boons to the productivity of both PM and of maintenance personnel. The shared PM divides the tasks between production and maintenance.

From a Lean point of view, TPM offers two opportunities. One is time savings, because the operator is usually already in close proximity to the machine, so travel time is reduced. The operator is also already in control of the machine, so custody transfer is simplified. The second opportunity is that TPM tasks feature lower skills (lower than full-blown maintenance activity) and are also usually a skill upgrade for the operator rather than a skill downgrade for maintenance professionals.

Several things are considered when dividing the tasks. Frequency is the primary divider. If the task is more often than monthly it is usually considered for the TPM team (preferably daily, or at most weekly). Of the frequent tasks, the ones that have readily-trainable skills are the best candidates for transfer to the TPM team.

3. Contrary to some interpretations, TPM is not primarily a maintenance program. It is either an operations program or a company-wide initiative. TPM must be implemented by all departments including maintenance, engineering, tool/die design, and operations, etc. Like many other programs of this type, TPM is a partnership between maintenance, production, and others in the company. The partnership will affect all the other stake holders of maintenance. This involvement is necessary for TPM to thrive. Every employee must be involved in TPM from the workers on the floor to the president.

Core delivery system of TPM outcomes is the TPM TEAM

By focusing on teams rather than individuals, TPM multiplies its effectiveness. The workers do not operate in a vacuum. The TPM team is based in the area and can have members from the ranks of maintenance, machine operators, and set-up people. Other members in engineering, quality assurance, and even accounting or marketing can be added on an ad hoc basis. Members of the teams are given the training, resources, time, and downtime, needed to accomplish their studies and equipment reviews.

4. TPM gives significantly increased power to the operators. TPM is based on the promotion of PM as a motivational technique through autonomous maintenance groups (operators have greater involvement and say-so about equipment). TPM works only because the operators begin to own the equipment.

Chapter 16

As ownership spreads, autonomous maintenance becomes a reality. TPM works inside the culture of the organization, to transform the relationships of people to the output. This new power changes people's minds about what they can have an impact on.

Success with TPM involves addressing the three very basic requirements that prepare the soil for the transplantation of TPM concepts and attitudes: These requirements are:

Motivation: The whole staff and all the workers need to be open to a change of attitude toward waste. This change is the same change that is needed for Lean principles to take hold, and it takes place over a period of time as organizations present, train, and start to change toward TPM. Motivation, in W.E. Deming's words, requires a consistency of purpose from top management because TPM is a long-term cultural change.

Competency: Certain skills are necessary before TPM can succeed. Training operators in PM, and design engineers and mechanics in root failure analysis, will eliminate waste and losses. One assumption of TPM is that operators have mental capabilities that are not being used in their current jobs. With training, these capabilities are gradually brought into use.

Environment: the top managers in the organization must support the thrust for improvement. The top people must understand the need for and the implementation of TPM. Top management must cultivate a trust of the shop-floor people. This trust is vital because there will be breakdowns and mistakes in the initial stages. Implementation of TPN is a long-term project so the operating environment will shift, and that shift has to be husbanded by top management, working with shop floor leadership.

TLC tasks are the Key to TPM

TLC (Tighten, lubricate, clean)

I re-coined the meaning of TLC to apply it to TPM activity. TLC normally means Tender Loving Care. When we apply this thought to machinery we get: Tighten, Lubricate, Clean. **TLC is the simplest way to reduce breakdowns, and is the bulk of TPM activity.**

One company found that 60% of its breakdowns could be traced back to faulty bolting (missing fasteners, loose or misapplied bolts). Another examined all its bolts and nuts and found 1091 out of 2273 were loose, missing, or otherwise defective. The JIPE (Japanese Institute of Plant Engineering) commissioned a study that showed 53% of failures in equipment could be traced back to dirt, contamination, or bolting problems. Effective TLC can impact other costs. One firm reduced electrical usage by 5% through effective lubrication control.

Tighten TLC

Bolting

"Bolts are tightened by applying torque to the head or nut, which causes the bolt to stretch". (See *Machinery's Handbook* published by Industrial Press). Good bolting practice takes a while to teach and is not necessarily intuitive.

Misconceptions

Using a torque wrench in infallible: Failure is not always because of friction. Remember, the goal is to stretch the bolt, and it is this stretching that clamps the joint. If there is rust or dirt on the bolt or nut, the torque displayed by the wrench dial will be greater than is being experienced by the bolt, so for the same amount of stretch you'll need greater torque. If grease is present, the torque required will be lower to achieve the same elongation.

It doesn't matter what the joint looks like when you pick a torque setting. Different joints require different amounts of torque. A joint in tension requires a different torque setting than a joint in shear. A joint in compression has significantly less torque requirements than either of the others.

All bolts of a particular size should be torqued the same. Bolts come in grades. The range of strengths between a grade 1 and a grade 8 bolt is almost 4 to 1. That difference may mean that the torque to stretch the bolt could vary as much (depending on the application).

Once you properly torque the bolt you're done. A well-known problem in mobile equipment is that bolts loosen up in the first 500 miles or 25 hours, and should be re-torqued.

There is no problem with a missing bolt if the others are intact. Loose or missing bolts are a major source of breakdown. Even a single missing or loose bolt might cause a failure. Although properly-engineered joints are designed with structural redundancy, each fastener is important.

Idea for action: The easiest technique to see if a nut has loosened is to scribe a line on the nut and the machine frame in a unbroken line when the nut is tightened correctly. This scribed line will stay intact (as a single line) as long as the nut doesn't move.

Lubrication TLC

Failures to lubricate are always the result of several factors. A leading factor is poorly- designed or -installed equipment where the lubrication points are too hard to get to, or there are just too many points. Other factors include use of too many different lubricants, not enough time allowed, contamination of lubricant or of the lubrication point, lack of standards, and lack of motivation of the worker. Don't assume that journeymen mechanics are experts in lubrication. In the US there are trainings and certifications for lubrication expertise.

Idea for action: Teach the oiler to follow the system of clean—grease—then clean again to avoid contamination.

Mistakes

Mistakes in lubrication can be devastating. The wrong lubricant could be spread to all the machines in an area in one afternoon as the lubrication route is completed. Too many choices in the plant lead to mistakes and supply problems. A better solution frequently, is to standardize on the 'better' product. That strategy will allow you to buy in larger quantities and reduce the price or increase the greasing intervals (because of the better quality).

Lubrication audit (partially adapted from <u>TPM Development Program</u>)

Good Practices Audit leads to Lean Lubrication
1. Are lubricant containers always capped?
2. Are the same containers always used for the same lubricants, and are they properly labeled?
3. Is the lubrication storage area clean?
4. Are adequate stocks maintained?
5. Is the stock area adequate in size, lighting, and handling equipment for the amount stored?
6. Is there an excellent long-term relationship with the lubricant vendor?
7. Does the vendor's sales force make useful suggestions
8. Is there an adequate specification for frequency and amount of lubricant?
9. Are there <u>pictures</u> on equipment to show how, with what, and where to lubricate and clean?
10. Are all zerk fittings, cups, and reservoirs, filled, clean, and in good working order?
11. Are all automated lubrication systems in good working order right now?
12. Are all automated lubrication systems on PM task lists for cleaning, refilling, inspection?
13. Do you have evidence that the lubrication frequency and quantity is correct?
14. Is oil analysis used where appropriate?

Problems that span lubrication and cleaning: Clogged, dirty, or broken lubrication fittings compromise the whole effort. Initial TPM cleaning should highlight these issues and correct them if necessary.

Chapter 16

Lean up your lubrication use

1. The strategy is to use oil analysis to see if the oil is still in good shape. Frequently, analysis can dramatically extend drain intervals without putting equipment at risk.
2. The second part of the equation is to either mount a bypass low-micron filter on the equipment, or purchase a filter cart and periodically (and thoroughly) clean the oil in place.
3. Look at the delivery of the lubricants to the place where they are needed. Transit in containers is a great source of contamination and of mistakes. In a truck fleet garage the ultimate motor oil delivery system is to pipe the oil right to the point of use, the repair bays.
4. Reduce the number of lubricants used, standardize on the best ones, and increase the quantity purchased.

TLC Cleaning

Cleaning is an essential activity that no one ever wants to do, but clean equipment provides several advantages over dirty equipment. On clean equipment it is easier to see leaks and small problems developing. On dirty equipment the leaks blend in with the dirt and even major deterioration is tough to see. The other issue is that clean equipment invites the TPM operator to put his/her hands on it, which makes for better inspections (TPM inspections are always hands on).

From the maintenance point of view, clean equipment is faster to repair when it does break. It is also safer to work on. Psychologically, clean areas are more attractive to get into and do not repel the maintenance worker (who might try to avoid interaction).

In a major cement-manufacturing plant, one of the engineers told the story that there was a major push to keep the equipment clean. After a few weeks the program was paying off and the place looked a lot better. Then, as he was walking through the area adjacent to the rolling mill he saw that the shell (outside) of the mill had a particularly bad-looking crack. The mill was taken out of service and the crack was repaired. The engineer speculated that, if the crack had traveled through the shell it could have caused an extensive downtime incident and could have resulted in a serious accident.

Cleaning Program Checklist

1. Cleaning main body of machine, checking and tightening bolts.
2. Cleaning ancillary equipment, checking and tightening bolts.
3. Cleaning lubrication areas before performing lubrication.
4. Cleaning around equipment.
5. Treating the causes of dirt, dust, leaks, and contamination.
6. Improving access to hard-to-reach areas.
7. Developing cleaning standards.

Keep area clean

Keeping things clean is not only a PM issue. Cleanliness is important in rebuilds, major repairs, and even small repairs. Any mechanic in the business for any length of time can remember a perfect repair gone badly because of dirt. Housekeeping is also directly related to safety. Accidents love dirt!

Beware

One of the problems with Total Productive Maintenance is that we tend to dump it into operations. We don't bother to prepare the ground, and do the training, and prepare people, and then build a reporting structure that will manage it. We just dump it, and then it goes away and then nobody's doing it, and it is no longer Lean.

CHAPTER

17

Quality

Benjamin Franklin says: Take time for all things: great haste makes great waste.

The ultimate in fat maintenance is returning to a maintenance job to redo it. Quality is the beginning point for Leanness. Without quality, all other Lean initiatives are wasting mechanics' time, parts, and machine time. There are many approaches to quality. The one taken by the originators of Lean Maintenance was the one promulgated by W.E. Deming.

There are excellent books on quality by and about Deming. He went to Japan in 1950 as part of the Marshall Plan. He taught many lessons, and the leaders of Japanese industry were receptive to them. For our purposes there were a few that directly lead to quality maintenance work.

Deming made 14 points in the training sessions that he led. These 14 points became the bedrock of an entire quality movement. He also made one observation about the morale of all workers (#13) that applies especially to maintenance workers. He said "make sure nothing stands between the worker and the feeling of pride in a job well done." This pride can be a motivator for both quality and Lean initiatives. Deming's 14 points were (with my apologies to the late Dr. Deming I adapted his points to the realities of maintenance):

1. Create constancy of purpose toward improvement of product and services. Large assets take some time to deteriorate. Only a department with a long view and a consistency of purpose will

catch the problems. Maintenance is no place for short-term views (although this is a very popular position to take right now). This constancy of purpose flows below any surface of management systems or choices. It is the same flow that keeps Lean Maintenance in place, healthy, and well.

2. Wake up! Awaken to the challenge. Take responsibility for and leadership in the Lean program. Don't wait until it is too late. Don't get too comfortable. Maintenance professionals must wake up; re-engage our world, and start thinking and using our creativity.

3. In maintenance and machine repair, inspection of work completed is important. But in itself, inspection will never ensure quality. If you catch a defect by inspection you will still have wasted effort to get that far. Having wasted time is already Fat. Cease dependence on inspection to achieve quality. Build quality into the process. How to build quality into maintenance processes will be discussed at length in another section.

4. There are hundreds of horror stories of selecting contractors or vendors based on the low bid. The practice of awarding business on the basis of price alone must be ended. Low bid is Lean, but constant problems with low-bid contractors is Fat. There is a major difference in rebuilding something like a major conveyor system if the contractor takes every care or just slaps the job together. The contractor's attitude can be the difference between a low bid and a middle bid. In some areas, the absolute cost of a job is important, but in maintenance, the asset will be with you for a while, so quality becomes far more important. Whatever the quality, the job you get might be with you for years or decades. Instead of low bid, minimize the total cost. Move toward long-term relationships of loyalty and trust. Allow your contractors to make a reasonable profit.

5. Constantly improve the system of maintenance and service, to improve quality and productivity, and thus constantly reduce costs. This reference is to the system that delivers maintenance. The system is like a river. Maintenance jobs are like boats. The jobs are important, and we want a fast-moving river, but not so fast as to smash up the boats. Rivers that are too slow will allow

the boats (maintenance jobs) to get stuck in the mud and then it will take a heroic effort to get downstream.

6. Institute training on the job. Some 85% of all maintenance knowledge is transmitted on the job, which means you are in the training business whether you like it or not. You might as well study training as a skill and get good at it.

7. Related to item number 6 is the need to develop the workers. Development and training are similar but not the same. Employee development is dedicated to enabling all persons to reach their own personal best (which we hope would be in service to the company). Institute a vigorous program of education and self improvement.

8. Leadership makes a major difference. Maintenance workers will follow leaders and keep working even when the boss is not around. The supervisor is the leader of that maintenance work unit. The aim of supervision should be to help people and machines do a better job. Supervisors should concentrate on eliminating any roadblocks that can interfere with maintenance work. They should be ahead of the action to anticipate and resolve difficulties before they become real problems.

9. It seems that companies use layoffs as the easy way to balance any reductions in work load with equal reductions in expenses. Consider any other alternative before using layoffs. Drive out fear of job loss, so that everyone can work effectively for the company. Companies should have savings gathered during busy times to fund projects that are essential for improvement during the lean times.

10. Human enterprise seems to grow silos or power centers. These centers are maintained by restricting the flow of information. Work consciously to break down the barriers between departments. Everyone's expertise is needed for Lean Maintenance and constant improvement.

11. The bulk of the problems with quality and production belong to the system, not the people. Therefore, eliminate slogans, exhortations, and targets for the workforce asking for such things as zero defects, or new levels of production. Such exhortations

create adversarial relationships and do not address the real issue—the business system.

12. Eliminate work standards or quotas, because the speed of work is regulated by the system. One of the goals of Lean Maintenance is to Lean down the business system that runs maintenance. Leaner systems will lead naturally to reduced time, with the workers getting the maintenance jobs done while working less hard.

13. Remove the barriers that rob the worker or engineer of his/her right of pride of workmanship. The responsibility of supervisors must be shifted from numbers to quality and improvement.

14. Put everyone in the organization to work to accomplish the transformation. This transformation is everyone's job.

How do we build a business system to reliably produce good quality maintenance work? A reliable system might require a less lean staff in exchange for fewer quality problems and more efficient job execution. Herein is one of the pitfalls of superficial approaches to Lean Maintenance.

Pitfall in Lean Maintenance

Organizations cut out middle management as a way to cut costs, flatten the organization, and simulate getting Lean. A Lean organization's goal is to have very few steps between the shop floor (where value is being added to the product) and the executives. There is no intrinsic problem with this approach.

Where companies fail is when the middle managers (who were just eliminated) were providing functions that are now not being provided by anyone. You cannot cut functions without impacting performance unless you also redesign the process. The new process must either re-allocate the task to someone else, have a technology perform the task, or redesign the process to eliminate the task.

Some of these functions (like planning and scheduling, as discussed in a prior section) can save 5 or more times the salary being eliminated. Elimination of the person in that key role will actually cost the company more money than keeping him/her. The functions that directly impact quality concern the missing elements of maintenance planning and scheduling.

Effective maintenance work requires that hundreds of details are coordinated in time and space. Planning identifies all the details, and scheduling moves those details to a particular time and place. The first step is to ask is, what are the elements of maintenance work that have to be present for the job to have the highest probability of high quality? These elements are:

1. Person(s) with the right skill(s) to perform the job that is physically able and mentally alert.
2. Correct parts, materials, supplies, consumables for the job.
3. Correct tools.
4. Adequate equipment for lifting, bending, drilling, welding, etc.
5. Personal Protective Equipment (PPE).
6. Proper permits and Lock Outs and therefore custody and control of the asset.
7. Safe access to assets, safe work platforms, and humane working conditions.
8. Up to date drawings and wiring diagrams and other information.
9. Proper waste disposal.

Planning is the process of identifying all these items (and some other more esoteric ones too). When the cost cutters eliminate the planning function they throw the workers out into the field without anyone having looked at the jobs. Some of the remaining workers are fine with this, and step up to do their own planning as well as they can, in limited time and without support. Other workers founder, and quality and productivity suffer.

Scheduling is the activity of taking all the items uncovered in the planning process and making sure they show up at the right place at the right time. Without scheduling, jobs would be started without materials. When materials are not present the workers might take it upon themselves to improvise. Improvisation might be great in the theater or in politics, but it is potentially deadly, and can certainly compromise quality in maintenance.

Contribution of 5S to Lean Maintenance

Benjamin Franklin says: A place for everything, everything in its place.

The system known as 5S is another transplant from Japan. Its basic function is to clean up and Lean up the workplace. Traditional Lean manufacturing uses 5S as a way to uncover problems in shop arrangement (for efficient production), and to make sure that what is needed to do a job is near at hand. In its original form, 5S was never intended for maintenance. It has been transposed into maintenance because the things that make it valuable in production are the same things that make it valuable in maintenance.

Studies of productivity of maintenance workers in heavy industry show the average maintenance workers are spending 27% of their day finding and organizing tools and materials, and transporting them to the job (the highest areas of losses). Great savings can be obtained from having what is needed at hand, but that requires knowing what maintenance job is going to be running ahead of time, or designing a multi-use cart or truck to take to jobs. Before we deal with those specifics, let's look closely at what 5S entails.

The 5S process is one of the Japanese systems that came over, along with JIT (Just in Time) manufacturing, with the TPS (Toyota Production System). The roots of 5S and Lean Manufacturing are the same. By removing the clutter and allowing you to see what is there, 5S techniques will uncover great opportunities for Lean projects.

Chapter 18

Wikipedia says "The assertion of 5S is, by assigning everything a location, time is not wasted in looking for things." This attitude is certainly in keeping with the goals and processes of Lean Maintenance. "Additionally, it is quickly obvious when something is missing from its designated location." In interviews with maintenance professionals, one of the most frustrating aspects reported is not being able to find critical tools and spares. As mentioned earlier, studies also show that this area has the largest pool of lost time in maintenance. Use of 5S provides the discipline and management structure to keep things in order.

Because the words are in Japanese, we see different translations of the 5 activities listed, usually to suit the environment where they are being used. Continuing from Wikipedia, one version of the translation of the 5S's is:

- Seiri: Separating. Refers to the practice of going through all the tools, materials, etc., in the work area, and keeping only the essential items. Everything else is stored or discarded. This process leads to fewer hazards and less clutter to interfere with productive work.
- Seiton: Sorting. Focuses on the need for an orderly workplace. "Orderly" in this sense means arranging the tools and equipment in a sequence that promotes work flow. Tools and equipment should be kept where they will be used, and the process should be ordered in a manner that eliminates extra motion.
- Seis: Shine. Indicates the need to keep the workplace clean as well as neat. Cleaning in Japanese companies is a daily activity. At the end of each shift, the work area is cleaned up and everything is restored to its place. The key point is that maintaining cleanliness should be part of the daily work—not an occasional activity that is initiated when things get too messy.
- Seiketsu: Standardizing. Refers to standardized work practices, but to more than standardized cleanliness (otherwise this S would mean essentially the same as "systemized cleanliness"). This S means operating in a consistent and standardized fashion, in which everyone knows exactly what are his or her responsibilities.

Contribution of 5S to Lean Maintenance

- Shitsuke: Sustaining. Refers to maintaining standards. Once the previous 4S's have been established they become the new way to operate. Workers must maintain the focus on this new way of operating, and not allow a gradual return to the old ways of operating.

In typical manufacturing facilities, little thought goes into the layout of the work benches and rebuild areas beyond where to put them. The photograph shows a precision construction equipment transmission and engine rebuild area. To move in anything to be worked on would require the technician to clear an area and move the junk around. He/she would then have to root around to find tools, spares, aids (such as scribers and layout fluid), and space to lay things out.

A 5S project in this area would perhaps have the following activities performed by the people that work there and extra help as needed.
- Clean the whole shop completely and paint walls and ceiling.
- Deep clean, paint, and seal floor (helps the lighting also).
- Increase the lighting in the work area and make the lighting more user friendly (color and type for tasks).
- Have re-builders lay out the shop for the way they would like to work and discuss, design, and optimize locations for tools, work areas, consumables, printed matter, etc.

- Put up wall shadow boards for common shop tools (not personal tools).
- Obtain drawer cabinets with dividers, and put away small parts near where they are needed.
- Clean machines and repair any leaks (inspect and PM machines and tools while cleaning).
- Paint machines.
- Look it over, does it scream efficiency?
- The whole job should cost less than a few thousand dollars.

How to keep the maintenance shop clean after the first blitz:

- Set up standards and post them: standards should define responsibilities for cleaning, identify proper methods and tools for cleaning, display and have proper PPE available for the persons doing the cleaning (gloves, face-masks).
- Establish time every day for cleaning.
- Randomly inspect.
- Identify the sources of dirt and try to eliminate them, and reduce the need to clean.

There should be quick paybacks in the form of faster rebuilds and more importantly, better quality rebuilds.

The most important of the 5S's is the last one. In a recent newsletter Imants BVBA (a European consulting companies' 2007 Newsletter issue) the last step of 5S is discussed in depth. This last 'S' is Shitsuke, meaning 'Discipline', and it must ensure that the first four S's are used properly. Discipline is essential for any program to be successful. It is the structure that reminds people that this is what we do—we put things away after using them.

Discipline means that everyone understands, obeys, and practices the rules. It encourages the good habits of 5S (while getting rid of the bad ones). One of the issues is the ongoing discipline to keep the place in order and maintain the improvements. Discipline also means that people are looking for opportunities to improve the order and flow.

Contribution of 5S to Lean Maintenance

Discipline can be achieved by:

- education
- learning on the job
- making job descriptions, work instructions, and schedules
- performing audits using checklists
- modeling the behavior that you desire from your employees
- daily management attention and interest

Keep in mind: 5S requires consistency of attention and effort. To make 5S work, it is critical that performance be measured, and that top management is committed to support the program.

What is the balance between 5S and Lean? The 5S system can be costly, and the benefits are limited. You can sweep up a maintenance shop, or you could clean it with a toothbrush. In other words, you can go too far. The balance is not easy to see. Always think through the effort level and the expected benefits, and pick the level of organization and cleanliness that works for your situation. The first 50–60% of the effort will result in 90% of the benefit. So it is important to get the place cleaned up and organized. The prospect of further benefits might make it worth going to the next level, particularly in larger or more complex shops.

CHAPTER

19

The Lean Machine

The original choice of the machine is important for running a Lean operation. Lean machines do the job day in and day out for as long as they are needed. They don't need much input so far as maintenance resources are concerned. Although they might not be the cheapest machines, they cost no more than they have to, given their function. These machines have no unused systems or capabilities.

Poorly-selected equipment can almost never be made Lean without complete re-engineering and/ or complete renewal. We can run Lean projects around fat machines, but there will be a limit to our effectiveness. Lean machines (and all other assets) can be viewed from several perspectives. One perspective is the choice of specifications that support Lean operation and maintenance. These specifications could include 'better' choices for components such as bearings, controls, mechanical strategies, and software.

There are machines that are more or less Lean from a manufacturing point of view. There are also machines that are Lean from a maintenance point of view. In this context we need to have machines that are Lean from both points of view.

This Lean journey can take several years. George Koenigsaecker, President of Lean Investments, LLC., tells the story of making Lean machines at Hon manufacturing. "One of the things that is unusual about Lean is that every time you reapply the tools and concepts to a given work area, you will identify new levels of waste and make new improvements. You will not be substantially "lean" until you have restudied every process—both production and administrative—about six to eight times."

The Lean Machine

At HON, even with someone who knew where we should be going and had a guideline for machine design, we were not able to "go directly there." It took several years of hard lean-conversion experience before folks thought that the stuff might really make sense. At that point, we started to do machine-modification work in addition to the setup reduction that we did early on. Our largest business involved a lot of sheet metal work, and we had lots and lots of press brakes. One of our first efforts sought to right-size a press brake.

Most of our brakes were for work pieces up to 12 or 16 feet long. Our earliest efforts at lean involved putting multiple sets of tooling in really long brakes in an attempt to reduce setups. Of course, really long brakes are great barriers to flow. So we decided we would get "right-sized" press brakes.

Year One: We bought a commercially available six-foot long press brake—an appropriate first step.

Year Two: Our newly formed machine-design-and-build group designed a three-foot press brake. A good step. And we were working to incorporate setup reduction concepts (robust design) and hanedashi. But we realized that we made many parts that did not require a machine of this size.

Year Three: We designed a two-foot-long press brake, and began serial production of them. The sequence was starting to get interesting. But as we looked at the design, we realized we had employed expensive, more-complex, variable-pressure hydraulics "just in case."

Year Four: We obtained simple hydraulics that only had enough power to make a part in ½ takt time, at its fastest speed.

Year Five: We finally "got it." On reflection, I realized that my guidance had been wrong. I told my design-and-build team that we no longer wanted to design right-sized press brakes. After four years of trying to do so, folks wondered where this effort was going. The new direction was to design self-actuating tooling (special machines) that met the key design criteria. At that point our creativity exploded, and we found many ways to make tooling work—from air bladders to hand-cranked presses for really small parts, etc.

Basically it took us eight years to understand the lean-machine concept well enough to consistently produce such equipment. So far it has not been possible to get machine builders to head down this path. Probably it takes so long because it goes against several fundamental machine concept paradigms that they hold dear.

Another approach is toward standardization. Maintenance departments have investments in spares, tools, vendor relationships, and training, so keeping the same manufacturers is Lean (in the absence of significant advantages from a change of OEM vendor).

I was working on automation of a through-putting oil terminal (a terminal that took oil off the pipeline or off ships, put it into tankage, and then dispensed it to individual trucks. The company was a service provider and did not own the oil.). We had discussions about the types of controls needed to activate the motors and valves. The terminal manager told us that we could put in any kind of controls we wanted so long as he could get spares from Square D (in other words only Square D controls).

He said he had spares, test gear and expertise with Square D and didn't want to duplicate that for a second vendor. He also said he had a relationship with the local distributor that was extremely important. I didn't know what that meant until in the middle of the night one winter, we had a problem with a programmable logic controller (PLC) on one of the machines. We needed a spare power supply (this was the second time it had failed in a short period). The terminal manager called the distributor at home, and within an hour we had the replacement power supply. The next day we had an engineer on-site, looking at the application. The engineer made some recommendations that eliminated the root cause of the problem, and we never had that issue again.

You can go a long way toward Lean with proper specification of the 'right' components for your use and your environment. I was visiting a small metal-stamping factory outside Philadelphia and asked about their attitude toward Lean machines. This was a small shop with very high levels of automation. Superficially it did not look in the least bit Lean. It was messy, and things were certainly not in their places.

Most of the machines and automation were designed and built in-house. When they needed a new machine they would take a used machine, strip it down, and re-build it so that it operated at double or triple the original speeds. They often built special machines from scratch because they could find nothing to perform the actions that were important to them. They built automation that would load and unload the machines, and allow one person to run several machines. The chief engineer said that was the only way to make money in competition with off-shore producers with low labor rates.

First, they said the machines have to produce the product within the quality requirements at the speeds necessary to make money. Second, they were looking for machines that spent the bulk of their time actually producing. These requirements seemed so simple.

I asked the chief engineer how they achieved these goals. He smiled and said that, when a machine breaks a second time (in a similar failure mode), he redesigns the machine to never break that way again. Sometimes this sequence took a few generations. But in the 30 or so years they've been operating with this philosophy, they've designed out many of the common maintenance problems.

In fact this firm operates without a formal maintenance effort, though they do have people who can do maintenance when necessary. These maintainers are the machine and tool builders themselves. Machine operators perform TLC (basic maintenance functions) as part of their jobs. Everyone on the floor seems to be sensitized, so that when the infrequent breakdowns do occur; it's something of an event. As often as not, the nearby toolmakers, engineers, and even operators, converge on the machine to figure out the root cause and discuss a permanent fix for the problem.

This plant is not Lean in the conventional sense, or at least it doesn't look that way. On the other hand, for 30 years they have competed on an even footing with the lowest-cost producers in the world (and managed to make money doing it).

The success strategy of this company has to do directly with their understanding of Lean machines. They invest in the machines where the investment will make a difference, and they decline to invest in any area that will not impact the final product. Their machines are efficient from all perspectives. Industry's attitude toward maintenance might some day catch up with them.

Their attitude can be summed up as: "Are failure and cost histories available to designers or equipment buyers to make better decisions?"

Chapter 19

Yes, because we already have a history that could guide people to specify better designs, based on the disasters of the past few designs.

One of the simplest examples of this attitude was an analysis done at a school district in Florida. This school district studied repair and maintenance of their various door systems. The district was lucky because they had pretty good data for almost 15 years about the repairs and maintenance done to the doors, and they had good records concerning the types of doors that were installed. In Florida, most doors open to the outdoors, and the county is located within a few miles of an ocean environment.

The first part of the analysis concerned installation of the doors. It was found that when properly-installed (meaning that the doors were installed so that they could swing 180° (all the way open, and there were frictional or magnetic catches to hold them open), the doors had significantly-fewer service calls. The results of these investigations led the group to develop a district-wide installation specification.

The district uses three types of doors, of which the cheapest were wood-core type. The costs of steel doors were in the middle, and the most expensive were fiberglass-clad doors with aluminum frames. When maintenance, repairs, and painting were taken into account, there was an interesting result. The most expensive doors needed almost no maintenance for the 15-year period. They also still looked presentable after years of service. The cheapest wood doors needed, on average, annual paint and services for various problems, and they looked worn after only a few years. Thus the aggregate costs of wood doors were by far the highest. The steel doors were in the middle with frequent painting, and a lot of complaints about the weight, and kids not being able to open them.

The study made it clear that the most expensive doors were the cheapest in life-cycle cost, and pleased the customers most. The payback of the extra cost was about 5 years longer than for the wood doors.

The approach discussed so far has been intelligent specification. Another approach is strictly economic. You can make a decision based on data and have an impact on Leanness for the next 20 years. Lean exists in both physical space (the machine or asset itself) and economic space (the money spent every year). Most people would argue that all Lean decisions are ultimately economic ones. This argument might be true, but looking at the problem from different perspectives has enormous value, and tells more of the whole story. The economic approach is essential nevertheless.

When we scale up from individual machines to whole processes, or even entire plants, lean design becomes essential. Two questions will highlight the Lean attitude on the part of the designers. These questions hinge on the maintenance department taking on the role of experts in maintainable designs. In other words, are the maintenance professionals experts in not only maintenance but also in maintenance-free designs? Expertise is also needed in looking at designs that, when they do need maintenance, the maintenance effort is efficient (leave enough space, install lifting gear, fall-protection, attachment points, and so on). Finally, positive answers to the questions also hinge on the rest of the organization viewing maintenance professionals as experts in other areas Questions that may be asked include:

1. Are drawings and specifications on all new processes and buildings reviewed by the maintenance department early enough in the process that changes can be made without adverse impact on the whole project?
2. Are specifications for buildings and new technologies on machines and building systems discussed with the maintenance department to see if they are in line with existing skills, training, parts stocked, test equipment, and tools.
3. Is there an ongoing process to develop a set of specifications and machine best practices? Of course there will always be disruptive technology, and we want to be the last group to stand in the way of better and potentially Leaner ways of making products.

The first thing that has to be overcome for Lean machines is the limited attitude of organizations toward acquisition costs. Buying machines on low bid will rarely result in a Lean environment. Pushing a project engineer to meet an arbitrary budget (that was pulled out of a hat like a magician's rabbit) will rarely result in a Lean machine. Lean machines need time, thought, and money. Of course we are also not talking about unlimited budgets. In fact, the quality guru identified inadequate time to commission new machines and processes as a barrier to quality.

There is a well-understood way to approach design or acquisition that will result in Lean machines. We are talking about looking at the life-cycle cost of the asset, where acquisition costs are just one part.

Another issue is how you institutionalize change. The human tendency is to fall back to older patterns. The important thing is to make the new change structurally. Make some real changes on the floor, flow, or process, to anchor the new reality. For example, if you make reliability improvements you have to reduce buffer inventory to maximize the savings. Any changes would cause a trickling through of the production to the MRP. Changing the buffer institutionalizes the changes and prevents you from slipping backward.

Life Cycle Costs

There are five cost areas that contribute to Life Cycle Costs. The life cycle cost is the total of all the five cost areas for the life of the unit. In overall financial terms, **the life cycle cost should be the determining factor in machine selection.**

Life cycle costs are cost projections so they are guesses about the future of energy prices, labor, maintenance costs, and other factors. In a complete analysis, the production rate is a denominator. In other words, the life cost per piece or per unit of output is the important number.

If it is assumed that two machines have the same output, there are two ways to evaluate life cycle costs, which are different only in the way they handle the time value of money. Method #1 disregards the time value of money and just looks at the estimates of the total costs:

Life Cycle Cost (LCC) = Ownership costs + Operation costs + Maintenance Costs + Allocation of Overhead Costs + Downtime Costs

The Lean Machine

The second method includes the time value of money and weights the investment by when it occurred. Breaking down the life cycle costs into specifics results in a deeper understanding of what contributes to the total cost of an asset and provides understanding toward a true Lean attitude.

Before anything happens you have to buy the asset. Sometimes money is invested long before any return comes back. This money goes for:
- Purchase costs, depreciation every year, and costs of money
- Lease/rental payments (fixed portion)
- Insurance costs, self-insurance reserves
- Permits, license costs, statutory costs (costs mandated by laws)
- Make-ready costs and installation costs
- Costs of searching, shopping, and bidding for machines
- Re-building and/re-manufacturing costs

It costs money to run a machine. These operating costs contribute to the Leanness of the process. Some of the operating costs might include:
- Energy and other utilities (water, compressed air)
- Consumables
- Operators
- Raw materials of all types

In manufacturing, the largest cost charged to maintenance (by far) is the downtime cost, or the money lost when the asset is not producing. In addition to the cost of lost production, there might be other costs including:
- Idle Operators
- Replacement or rental unit costs (such as renting a compressor to get back into production)
- Replacement costs for ruined products (or reprocessing costs for materials left in pipes when the plant went down)
- Late penalties
- Start-up costs
- Shutdown costs

- Intangible costs of customer dissatisfaction, hidden costs, other costs

We have been discussing above- and below-the-waterline costs (iceberg analogy). The maintenance costs labeled here are the ones above the waterline. All the other cost areas are below the waterline. Maintenance costs can be broken into:
- True cost of inside labor (including fringes, lost time, and overhead)
- True cost of inside parts (including cost to carry, spoilage, etc.)
- Outside Labor (contractor)
- Outside parts (vendors)
- Consumables to perform maintenance (such as welding rod, rags, etc)
- Costs of rental equipment
- Any hidden costs of failures (such as disruption to other processes)

Of course there are invisible or behind-the-scenes costs that are called overhead costs. These overheads cover the basic costs of running a business. They can include:
- Cost of Maintenance Facilities
- Heat, Light, Power, Phone
- All persons in maintenance department not reported on repair orders
- Supplies not charged to repair orders
- Tools and tool replacements
- Repair of Maintenance Facility, maintenance tools
- Clean-up
- Computer systems, all expenses

20

RCM and Lean Maintenance

RCM is problematic for Lean practitioners, probably because of how RCM is thought of and implemented. RCM projects are very expensive and intensive. Sometimes, critical people are taken off the shop floor for days at a time to work on RCM projects. On the plus side, RCM is the most rigorous method we have to analyze a maintenance event and come up with definitive answers about tasking, frequency, and redesign. On the negative side, RCM is only justifiable when there are large amounts of money involved, or where it is thought that safety or an environmental catastrophe is at stake.

Still RCM offers Lean Maintenance efforts a level of rigor not available in other programs. It also offers mental models of maintenance beyond anything else. We can learn a great deal from a systematic study of RCM. Keep in mind, RCM is one of the most powerful ways to improve maintenance because it addresses the core of the customer need, that is, an increasingly reliable system. The technology is an outgrowth of deep investigations into reliability that were performed on behalf of aircraft manufacturers and airlines.

As mentioned, RCM developed out of the aircraft and airline industries. One of the first airplanes that used the techniques from design to delivery was the Boeing 747, the first of which was delivered in 1969. The 747 was the first jumbo jet and was designed to replace the smaller, older, intercontinental 707 jet-liner.

A great deal of behind-the-scenes maintenance is done on aircraft. Periodic inspections are performed, components are overhauled, and even major sections of the aircraft are stripped and rebuilt. In fact the older 707 took 100 hours of behind-the-scenes maintenance to support a single hour of flying time. The larger and more complex 747 improved that benchmark dramatically to 10 hours of maintenance for every hour of flying.

The whole airplane system was measured in terms of incidents per 100,000 take-off and landing cycles. An incident had a specific definition of a failure of a major system, but not necessarily a crash. The 707 was pretty reliable, with a rating of 5.6 incidents per 100,000 cycles. The 747 had increased reliability of 1.6 incidents per 100,000 cycles. Imagine the results if you could reduce maintenance by a factor of 10 and increase reliability by more than 3. That is why people started to look closely at RCM as a way toward Lean Maintenance.

RCM is a five step process. The process is usually team-driven, with members from operations (pilots), engineering, and maintenance (where there is significant hazard, safety or environmental specialists would be included). The work is usually facilitated by a RCM specialist with good knowledge of the process and products.

Here is a (very) quick review of the RCM technique.

Steps in RCM

1. Identify all the functions of the component. At first this step might seem trivial. Functions are divided by the terms primary, secondary, and protective. Each function is defined by a specification or performance standard. The question you need to answer is, what is this item for? There might be several functions. If you look at a jet engine you might find functions including thrust, braking, electricity, hydraulic energy, compressed air, fuel containment, and that doesn't even consider air flow, sound reduction, and others.

Another example would be a conveyor in a quarry. The primary function of a conveyor is to move stone from the primary crusher to the secondary crusher. The specification might call for 750 tons per hour capacity. Secondary functions include containment of the crushed stone (you don't want pieces falling through the conveyor and hurting someone).

RCM and Lean Maintenance

Hidden failures

Certain assets have special safety functions. They include fuses (protection from over current), pressure relief valves (protection from high pressure), etc. These devices are called hidden safety devices. We also have to be conscious of the fact that some failures of sensors or protective devices are hidden. A failure is said to be hidden if it occurs and the operators, under normal conditions, would not notice the problem. For example, if a conveyor belt thickness gauge fails (unless the design is fail-safe), the operators would have no way of knowing that the sensor is out of service. After a hidden device has failed, a rip (of the belt) could develop and cause complete belt separation without notice, and thus cause an accident.

2. The second step is to look at all the ways the asset can lose functionality, called functional failures. One function can have several functional failure modes. One functional failure of the jet engine is no thrust; another could be partial such as a partial loss of thrust. Each of the functions has one or more functional failures. The hidden safety function can lose functionality by not working at all, by cutting out at too high a temperature, or by cutting out at too low a temperature.

For the conveyor example, a complete functional failure would be that the unit cannot move any stone to the secondary crusher. A second failure would be that it can move some amount less than the specification of 750 tons per hour. A third functional failure is if the conveyor starts moving more than 750 tons per hour and starts to over-fill the secondary crusher. Each secondary function also has losses of functionality. In our example, the conveyor could allow stones to fall to the ground, creating a safety hazard.

3. Review each loss of function and determine which of all the failure modes could cause the loss. In our example, the list might be 20 or more failure modes to describe the first functional failure alone. Particular care must be taken to include failure modes where there would loss of life or limb, or environmental damage. It is important to include failure modes beyond the normal wear and tear. Operator abuse, sabotage, inadequate

lubrication, improper maintenance procedure (re-assembly after service), all would be included.

Failure modes of our rock conveyor include motor failure, belt failure, pulley failure, inadvertently turning the unit off, power failure, etc. Each functional failure is looked at and the failure modes are defined.

Some judgment must be used to include all failure modes regarded as probable by the team. All failures that have happened in the past in this or similar installations would be included, as well as other probable occurrences. Take particular care to include failure modes where there would loss of life or limb, or environmental damage.

It is essential that the team identify the root cause of the failure and not the resultant cause. A motor might fail from a progressive loosening of the base bolts, which strains the bearings, causing failure. This event would be listed as motor failure, due to loosening of base bolts.

It is important to include failure modes beyond the normal wear and tear. Operator abuse, sabotage, inadequate lubrication, and improper maintenance procedure (re-assembly after service), would be considered, for instance.

    4. Assess the consequences of each failure mode. Consequences fall into 4 categories including safety, environmental damage, operational, and non-operational. A single failure mode might have consequences in several areas at the same time. John Moubray in his significant book *Reliability-Centered Maintenance II* says "Failure prevention has more to do with avoiding the consequences of failure then it has to do with preventing the failures themselves."

The consequences of each failure determine the intensity with which we pursue the next step. If the consequences include loss of life it is imperative that the failure mode be eliminated or reduced to improbability.

In the conveyor, a belt failure would have multiple consequences that would include safety and operational. A failed belt could dump stone through the conveyor superstructure, hurting anyone who was underneath or nearby. The failed belt would also shut down the secondary crusher, unless there was a back-up feed route.

RCM and Lean Maintenance

The failure of the drive motor on the conveyor will cause operational consequences. Operational consequences have costs to repair the failure itself, as well as the costs of downtime and eventual shut-down of the downstream crushers.

Other failures might have only non-operational consequences. Non-operational consequences include only the costs (all above the waterline), to repair the breakdown.

5. The final step is to find a task that is technically possible and makes sense, that will detect the condition before failure, or otherwise avoid the consequences. Where no task can be found, and there are safety or environmental consequences, then a redesign is demanded.

For example, if it is found that the conveyor belts start to fail after they are worn to 50% of their thickness, inspections, with belt replacement when needed, might be indicated. If the belts fail rapidly after cuts or other damage, then a sensor might catch these problems. Wherever safety or environmental damage is the main concern, the task must lower the probability of failure to a very low level, or mitigate the consequences of the failure. If the failure mode dumps 7-inch rocks on anyone standing under the belt, then installing steel liners at critical spots to catch the rocks might mitigate the problem.

Cost

Much of RCM is related to a comparison of the costs of failure and the costs of the PM task. In operational failure modes (such as the motor failing), the cost of any PM tasks over the long haul has to be lower than the cost of the repair and the downtime. If the PM tasks cost $1000 a year, breakdowns cost $2500, and downtime costs $4000 to repair, then the breakdown has to be avoided more than every 6 years. This consequence-driven maintenance activity is the core of the contribution RCM has made to the wider maintenance field. Since RCM concepts have been introduced, most conversations about failure discuss consequences. Ultimately RCM is a Lean Maintenance approach.

21

Lean and RCA (Root Cause Analysis)

The Leanest Maintenance is no maintenance. Root Cause Analysis (RCA) eliminates whole maintenance situations by eliminating the causes of those situations. Nothing is more lean than no problem in the first place.

Toward the end of the heating season one year, I received a call to say there was no heat in a single-family rental house I owned. I visited the house, examined the forced-air furnace, and found the heat exchanger was shot. I hired a local handyman to replace the unit. I did not receive another complaint so I forgot about the issue.

During the first cold snap the next winter I got a call from the family, again complaining about lack of heat. This time I sent a licensed HVAC contractor to look at the heater and repair it if necessary. He reported that the transformer was bad, and he had replaced it, and sent a bill for the service call. I didn't get another call from the family, so I assumed the problem was solved (the weather warmed up also). A week or two later I got another heat call. I asked the same contractor to take a look. Another technician looked the heater over and found the transformer had failed and now the thermostat also was not working.

Now I was annoyed and asked for some consideration since I thought the contractor must have used a bad transformer, which I didn't want to pay for twice. When the third complaint came in I called the contractor and asked him to meet me at the house. We found the thermostat and transformer blown again. We both agreed that didn't make any sense. When we tried to fire the unit with some new components in place it wouldn't fire.

Lean and RCA (Root Cause Analysis)

We both were perplexed and decided to do a more thorough examination. We found that the gas valve was wired incorrectly (presumably at the factory). Instead of the circuit being across the valve coil (and providing 300 ohms of resistance), the factory had wired the coil out of the circuit, and the result was a dead short in series with the transformer, thermostat (and safety switches). After a while, the points inside the thermostat would fuse, and eventually the transformer would be fried. And every time, the heater would stop working.

04-FEB-2007		**WORK ORDER REPORT**		Page 1 of 1
128088	BAD VLVE. ASSMBLY.			

Status Code:	INPRG	**Work Start Date:**		Parent:
Report Date:	04-FEB-2007	**Work Completion Date:**		Followup to:

Description:

Location:	10QL1593	ALLEN CASE SETTLER SOUTH	MCC #
Equipment:	E1593	ALLEN CASE SETTLER SOUTH	ID Tag

Old Eq#:	1593	Crew: SWING			Est. Duration	0
Supervisor	Assigned to	Work Type	Work Class	Downtime Req		Priority
DS	DON	CM	1-MAINTENANCE	1-IMMEDIAT		2

Value of parts issued as of: Feb 04 2007 04:04PM **124.18**

Job Plan:
.00

Comment _Solenoid working to slow_

Hours Worked .50   Work done on linebreak ___

Proc ___ Elec ___ B&R ___ Pkg ✓ Field ___ Waste ___ Fklt ___ Othr ___

_2-4-07_   _1754786_
Date Completed   Completed By / Emp#   Sanitation   Supervisor
                           Don Birking

> Work Order says that the solenoid was "too slow." Presumably the technician replaced it (by being a detective you can deduce that). The question not answered was why it was too slow?. Perhaps by looking deeper and finding the root cause, we could have dealt with this problem forever.

WOPRINT.SQT

Chapter 21

Now it was the contractor's turn to be annoyed at his men because neither of them had tested the unit. They just replaced the faulty parts and never asked questions about why the parts had failed in the first place. He apologized for his workers and cancelled the billing. I also felt stupid because it was obvious that the handyman had not fired the unit either, and I had never checked the work. But with all that said, what is the barrier to doing a root cause analysis? Although you probably don't have anyone that would ever take a short cut like that; how many times have you had a technician just swap out the bad part, and if the machine started working, never ask the question, why did it go wrong?

Root Cause Analysis is a structured process to get to the root of whatever is causing a problem. Finding the cause is the basis of good maintenance practices, and is part of any 'good' technician's tool kit. In some places, the consequences, cost, frequency, or schedule, do not allow a proper analysis. For example, with the heater, if the first technician had gotten the unit to fire, a full analysis, tracing all the circuits, and testing all assumptions, would not generally have been necessary.

The RCA process is very straightforward but requires some people that completely understand it, and the likely failure modes and relationships between different components of an assembly.

The first step is to clearly define the problem being examined. It is important to keep that definition tight, and as simple as possible. The complexity of the analysis goes up exponentially as you add other items to look at. The probability of forgetting where you are and getting confused is also present.

It is important to secure the scene (just as in a crime scene) and control variables. In the residence, having the failed components from the heater would have aided the analysis. Sometimes the first response confuses the problem and makes the whole situation worse. In fact, some of the worst results have come from the actions taken and not the initial problem itself. After the scene is stable, try to find out what exactly happened. Take a good look at the whole process. Ask basic questions like: what process parameter was changed, or what did the machine sound like (what noises was it making) before it went bad?

Lean and RCA (Root Cause Analysis)

Ask yourself, is this a huge, profit-altering issue or a routine pain in the neck breakdown? The bigger issues need a more formal RCA process to be sure not to introduce iatrogenic (problems caused by servicing) failures into the system. Routine problems can have more informal RCA actions. These informal actions should be part of almost every great maintenance repair. This analysis is also Lean.

After the important first steps are complete, the brain trust gets together and looks for causes. The problem is well defined and the initial conditions are known as well as they can be, so the next step is to look for what could cause the problem. The procedure is similar to the failure modes exercise done in Reliability Centered maintenance except that we are looking for a single likely cause, not all probable causes.

Some of the causes have predecessor causes. Follow each cause back to the beginning. This sequence will develop trees of causes. Keep working these causes back to their roots, and you should be able to follow the logic back through the tree to the source.

The final stage is when it becomes apparent that there is a complete causal tree for the event you are trying to avoid. You can then work with the team to develop a change or a fix that will interrupt the tree. If the tree is designed properly, the change will stop the final event (the breakdown) from taking place.

Software is available to help firms manage this process, but, as with all these processes, it is important to have some time to work the problems back to their roots. Repetitive maintenance is Fat and RCA to eliminate the root cause is Lean.

CHAPTER

22

Lean Outsourcing

Benjamin Franklin says: A penny saved is a penny earned

Outsourcing for Americans and the West means sending manufacturing jobs to China and technology jobs to India in order to save money. Programming in India and manufacturing in China is cheaper for now. However, the costs of both locations have been increasing faster than the labors rates in the West. For the maintenance department, outsourcing is hiring outside contractors to perform some or all of the maintenance needed.

In Lean Maintenance, outsourcing is neither good nor bad. The projects we will be developing can be carried out by contractors as well as by in-house labor. One advantage of in-house labor for Lean Projects is that the motivational level often is higher. There are situations where contractors hold the clear advantage. In other situations the in-house departments have a clear advantage.

Toward the end of the 1990s there was a popular business concept that stated that companies should focus on their core competencies and outsource the rest. Companies outsourced their IT departments, their human resources, and almost everything else. Some major hardware vendors don't make anything, and have no production facilities themselves. Everything is outsourced.

If maintenance expertise is not determined to be essential for success, just call a contractor. Most of the US military bases are maintained by contractors. The urge to outsource maintenance can be driven by a belief that it will save money, but usually it is driven by inadequate service levels, or the inability of the in-house parties to manage the function.

Institutional barriers

A few organizations have institutional barriers to productivity that are easier to get around by going outside. Institutional barriers include unusually generous vacation/sick pay, personal time off packages, short work days, longer than usual or more frequent breaks, excessive indirect responsibilities, burdensome labor contracts, aggressively unfriendly labor relations, to name a few. Organizations with these issues might very well save significant money from outsourcing maintenance. This is one time where we will be breaking the rule laid out in the beginning and discussing job reductions and Lean Maintenance. You'll see in the example below this conversation is about the business environment rather than the people in maintenance themselves.

One clear case of outsourcing that was mentioned earlier was a Lean Maintenance bonanza. At a large urban airport, maintenance was performed by city employees. The union had very strict guidelines about not mixing trades, and prohibited multi-skilling. The airport had manning rules that required a licensed electrician, a licensed elevator/escalator repair person, and a general mechanic for the baggage system (among others), to be on site whenever flights were expected, or whenever the terminals were open. Traditionally, the municipal workers union interpreted this rule to require at least 3 different positions, working for 20 hours or more a day.

The city went through an extensive process to outsource the facilities maintenance. The contractor that was chosen was required to submit a manning schedule. The schedule showed one person for all three jobs. The person was an elevator/escalator mechanic who also was a licensed electrician and had qualified on the baggage systems. The contractor argued successfully that on the off shifts it was more efficient to have a smaller crew because there were few support people. This airport (unlike some others) had never geared up for night time maintenance (which would have made the discussion quite a bit different). Replacing three people with one with all the skills needed was Lean.

Organizations with these barriers can reap some benefit from outsourcing. A typical contractor will get 1850 hours of work in a straight-time year. A typical maintenance operation in a large company in the Northeast would get closer to 1775 hours per year (with the South doing a few more hours). That spread was not enough for the contractor to make a living.

Some firms offer significantly more time off, and available hours can run on as low as 1450 hours per year. These firms need to watch out. Contractors can come in and re-hire all the same people at the same wage rate (but without the time off), and sell the labor back to the company 10 or more percent cheaper and still make a profit. That is why almost no military bases are maintained by federal employees today. All the work has been taken over by contractors. Frequently the contractors hire those same employees back into their old jobs. In this example too, hiring a contractor would be a Lean choice.

Generally, for organizations without these barriers, contracting to save money will either not really work, or will work for a very limited time (of course there are exceptions). Without an edge (like those barriers already mentioned) it is difficult to make departmental outsourcing work financially.

One mistake made by organizations seeking to outsource maintenance is looking only at the maintenance workload in the outsourcing quote. Maintenance departments do a lot more than maintenance. A typical department might do construction, installation, event set-up, office moving, sanctioned community work, pick-up and delivery, picture hanging, car starting, snow shoveling, fire brigade, special projects, competitive intelligence, buying used machines, auctions, disposal of spare parts, and the list can go on. The issue is that, when the contractor only bids maintenance work, all this other work becomes extra (sometimes at premium rates). If you outsource, be sure to add these items into the contractor side, or subtract them from the maintenance side.

There are excellent strategies to aid the use of contractors in a Lean way. Many companies hire contractors to supplement their regular crews instead of adding permanent personnel. This approach gives the advantage of expanding the crew when needed and contracting it also when needed. The second is to use contractors for peaks in your business cycle. You might need a 10% larger crew from January to March to get the line ready. Contractors are the Lean alternative, unless there is work for the extra people year round.

Certainly, spending money on lawyers instead of maintenance is Fat. But spending money on lawyers to prevent paying more money to lawyers or contractors is Lean. If you contract out construction or other one-off events, review the list below and see if you can do some anticipatory work to make the whole process Lean.

Tips to Avoid Claims. Tips from the people in the trenches to make contracting go smoother

1. Scope the job and view the scope as a contractor would. Define the work to be contracted carefully. The better the definition at the early stages, the better the job will go. Avoid loose specifications (on both materials and work to be done). Discuss the quality of materials needed, and the timing. Communicate your ideas to the contractor (How would you make sure they understand? Can you explicitly answer that there was there a meeting of the minds?)

2. Before awarding the contract, pre-qualify the contractors to be sure they are strong enough to do the job. On larger jobs, check out the finances, credit, insurance, and staff, and get bonds. Visit other jobs to see their quality, and call references. Negotiate and award the contract. Of course after signing, don't lose track of the contract documents (Keep a fair and complete set of contract documents).

3. Beware of and avoid, if possible, low-ball bids. Negotiate a schedule of extras, and make it as complete as possible. Work on the concept of no surprises. A common contractor ploy is to low-ball the bid to get the job, and flood the company with small extras. Add in clauses like "all extras not included in the original price have to be agreed to in writing prior to commencement of the work."

4. Be sure to spell out deduction clauses in the contract. Spell out what you will charge back and when you will charge it.

Examples would be debris removal, clean-up, inadequate crew size (in fixed input contracts), missing firm completion dates.

5. Negotiate cancellation clauses. You need to spell out how and why you can cancel the contract. This aspect is particularly important with multi-year housekeeping contracts. Otherwise you may find yourself stuck with a mechanic's lien over an inadequate job, after you did not make the final payment.

6. On ongoing service bids, avoid both too short a contract term and too long a contract term. If the term is too short, the contractor will chart excessively for mobilization costs. If the term is too long you might be stuck with a barely-adequate vendor, with no easy way to improve the situation.

7. Be as complete as possible about responsibilities, who supplies what, where to unload, site rules (safety, user contact, clean-up, security, keys etc.). There should be statements about how the site is to be left at the end of each work day. Who is responsible for locking up, barricades, traffic management, cleaning, and debris removal? In one contract we required the contractor to enter all work order information into the CMMS. The contractor protested, but the requirement was in the contract that was signed. The agreement should also state clearly, who is responsible for municipal permits and plans.

8. Inspect the contractor's insurance policies. Have an agreement about what happens when (if) the contractor damages your property (or worse yet damages a neighbor's property who then might sue you, or it might spoil a good relationship). Require an up-to-date certificate of insurance covering: General liability, Casualty (property damage), Workmen's compensation, Auto liability, and if they do the design, Malpractice, and Errors and Omissions. The smallest contractors (sole proprietors) don't need workmen's compensation (check with your attorney or your insurance professional).

9. Define performance and state clearly what a good job would look like. Add clauses like "all work is expected to be done in

a professional and workmanlike manner. All work will be in compliance with "applicable city building codes."

10. Prepare the area to be worked on. Remove as many items as possible to avoid breakage and theft. If indicated, isolate the area so that contractors have no reason to wander around. Provide agreed-upon utilities and lighting. Increases in theft levels sometimes correspond with increased contractor presence in the building.

11. Manage the contractor. It may be necessary to require clocking in and out of the contractor's employees. Keep a record of the job as it unfolds, and provide feedback. Perform frequent inspections, and document the results. Have a functional, planned, schedule, and compare progress with projections. Problems should be identified as early as possible, and action taken to resolve them.

12. Include agreements about when and amounts of payments to be made, etc. in the contract. Avoid sloppy record keeping. Require paid receipts to prove that subcontractors and material vendors have been paid. Get a release of all liens form signed before the last payment. If you don't, you could have paid off the general contractor and still be hit with liens from unpaid subcontractors. Consult with your legal department about lien laws in your state, and be sure you are covered.

13. Resolve disputes promptly. Avoid endless delay in resolution of disputes or you'll end up in court.

14. Evaluate each contractor on a regular basis for quality, service, cost, and fulfillment of contract terms. Write up a short narrative to put into the file about how the job went.

(We thank Ed Feldman P.E. for his insightful avoid list and ideas about the bid package, some material adapted from the San Francisco State University Manager's Bulletin, and some from The Landlord's Handbook)

23

Some Lean Tools

Every management technique has its favorite tools. Project Management has its Gantt charts and CPM diagrams, and 5S has its shadow boards (that show the outline of the tool when it is missing). Lean maintenance also has charts. But Lean Maintenance has an additional advantage. Lean Maintenance encompasses any technique that reduces the inputs to produce a positive maintenance outcome. All the tools of all the techniques can also be Lean Maintenance tools.

Circle game

The first tool discussed in the first chapter was the circle game. In Lean Manufacturing, a useful technique is to draw a circle on the floor and have a Lean team member sit there observing the operation for a few hours. Waste then becomes obvious, and the observer notes the waste and writes down any questions that come up. In particular the observer is trained to look for the wastes of manufacturing: over-production, wasted waiting time, wasted time moving people, tools or materials, wasted materials, waste in the inventory, and wasted motion. More will be said on these wastes in the next chapter.

The circle technique has to be somewhat re-thought because maintenance does not take place in one place as production does. With some creativity we can gain real intelligence. Maintenance can use the technique to observe specific jobs, to observe the maintenance shop or the foreman's office. Maintenance has massive waste compared to a typical production line so it could be easy for an observer to be overcome with all the opportunities in doing floor-level observation. The important thing is to pick something and observe it until you understand it. And then use the processes described in subsequent chapters to isolate, define, and eliminate the source of the waste.

Let us speculate about what we would see while sitting inside a circle in a typical maintenance shop. For a lot of the day we would see nothing, because everyone would be out on jobs. The most fruitful times would be in the morning and evening, or the beginning and end of the different shifts.

What would we see? Would there be an intentionality to all movement, or would there be drifting. Would we see people reviewing job packages and mentally preparing for the day, or not? Is the place friendly, productive, positive, energizing, or not? Observe everything, and make notes, and have some fun doing it. It goes without saying that, before you draw the circle, you will have a few meetings about what this experiment is about (the most important concept is that whatever is observed is not personal).

The same circle could be drawn near the parts window, the tool crib, and near computers, and different observations can be made in each place. Jobs can be observed. A variety of different people should do the observing.

Displays

Displays are real-time representations of a process, technique, or procedure, and are extremely powerful in conveying the status of a project without confusing detail. As such, displays are universal tools for Lean Maintenance initiatives. They can range from a thermometer showing Lean savings to date, to a control chart showing temperature variation in a room.

One of the most famous displays is the Gantt chart. Harry Laurence Gantt developed his chart in about 1914, while working on artillery shell production at the Philadelphia Frankford Arsenal. Gantt thought the workers would work better, and be more motivated to make the production numbers, if they could see how they were doing.

Chapter 23

The problem was that he had workers that couldn't read, couldn't read English and some could barely communicate with supervision.

Gantt devised a chart that showed the scheduled production for the shift and the actual production for that shift. At that time, his chart (now the Gantt chart) had nothing to do with project management. The worker (whether or not he/she could read English) could readily see where production was. The feedback was found to improve production. Gantt lived on until 1919, all the time thinking about and improving his displays. The Gantt chart we use today is a descendant of that original chart and still just shows how the project is doing compared with how the project was supposed to do.

In Lean projects, anything we are measuring can be converted into a display. A chart with energy usage by day for last month and for this month, will tell a powerful story. A display showing board feet produced could be a power motivator (especially when coupled with incentives). But even without incentives, the chart will show how the project is doing, which in itself, seems to motivate most people.

Pointer Lists

In this book we will be using pointer lists extensively to find waste. Maintenance is made up of thousands of details. Hidden in those details are all the patterns, all the problems, and all the 'bad actors' that are such a problem for maintenance departments. Somewhere locked in there is all the waste. Every minute wasted, every part spoiled, is in there if the data is accurate, sound, and complete. The way to unlock the waste is to ask the right questions of the data.

With most set-ups, on the first go-round, we don't even have to go to the data. When people are asked for their memories of the items on the lists, their memories are biased, but they still effectively point to waste.

The first step is to make the lists. The lists used here are not all the pointer lists that could be made by any means. These are the ones for which it is usually easiest to gather data, or guess information for (like the most repetitive jobs).

The second step is to boil down the lists in team conversations and pick out those few items from some of the lists (5 to 7 items), that appeal most to the people in the group. The rest of the items are not discarded but are kept for the next go-round. Different groups might find that different list items appeal to them.

Once the lists are boiled down, the team starts to decide which projects will reduce the waste represented by the pointer. The final agenda of the projects, based on the boiled lists, is the program to be worked on from this stage. The only development at this stage will be preliminary, with a project name and a few lines of explanation.

After the list of projects is written, one of the selected projects is chosen for a focused assault. Choice of the project is based on specific criteria.

The fun then begins, and the project is chosen for execution, executed, and written up. The write up is important as explanation and guidance for the whole process and for subsequent teams.

24

Where and How to Look for Waste

Benjamin Franklin says: Waste not want not.

Most of the waste that is easy to get to is visible. Just look for trash, dirt, excessive work in process, people hanging around, and the accumulation of almost anything. The problem, normalization of deviance, will be discussed in depth in a subsequent section, but it is the reason why it is hard to see the accumulations. The piles look normal, and that is where the pointer lists come in useful.

Waste reduction can be as simple as improving the kinds of paint brushes (to reduce the time it takes to paint something, reduce the amount of paint used, and improve the quality of the job). Reduction of waste could also require deep knowledge of pump applications, such as evaluating two different pumps in the same application (with consideration of initial cost, availability, energy costs, reliability, repair costs, availability of spare parts, downtime, and downstream impacts).

In traditional Lean Manufacturing, several waste areas are identified. These areas have analogies in Lean Maintenance, and some represent major losses.

- Over-production is one of the uncommon wastes in maintenance. An example is when you are rebuilding valves and you might rebuild too many for immediate use.
- Time wasted in waiting is a major loss in maintenance. We are always waiting for another trade, for operations, for drawings, for something. This waste is one of the items that can be reduced by multi-skilling, planning, and scheduling.

Where and How to Look for Waste

- Time wasted moving people, tools, or materials to and from jobs. This waste takes up over a quarter of the day of a typical maintenance person. Effective planning can reduce this waste by half.
- Wasted materials are a problem in maintenance and particularly in projects. Sometimes it is difficult to return surplus project materials or to find other uses for them.
- There is significant waste in the maintenance inventory, which is a ripe area for Lean work. We want to have just the materials we need, no more and no less.
- Wasted motion is a difficult area because each maintenance job is somewhat different and there is no universal solution. Major organizations such as UPS (the package delivery and Logistics Company) have studied maintenance work, and teach ways that waste less motion.
- The final waste is defects, or maintenance jobs that have to be reworked. This waste is the bane of the maintenance department. Proper tools, training, and parts, reduce this effect, but even vigilance will only reduce it, and nothing can completely eliminate it.

There are opportunities for cutting waste and making improvements in every maintenance operation. An internal study done by a major maintenance provider in Canada provided a way to quantify the opportunity. Details follow:

Percentage of possible savings of maintenance budget dollars

(Above-the-waterline costs)
- 39% Re-engineering of equipment, and maintenance improvements to equipment
- 26% PM improvement and correct application of PM
- 27% More extensive application of predictive maintenance
- 7% Improvements in the storeroom

Work in each of these areas can result in significant reductions in cost or improvements in operation. Before getting into specific areas of waste, let's take a look at the barriers to seeing waste.

Chapter 24

Normalization of Deviance

There is a psychological phenomenon that works against maintenance people with long tenure in the same plant. In fact, because of this phenomenon, maintenance professionals are dramatically better at seeing waste almost everywhere else than in their own plant. The psychological phrase is "Normalization of Deviance." Mike Mullane, retired Space Shuttle crew member, spoke to an IT group. His main message was that, too often, individuals and organizations give way to "normalization of deviance." That is, we accept results below our own specifications, and then wonder what happened when disaster strikes. The prime example for NASA was the loss of not one but two shuttles. The causes differed, but the acceptance of what should have been unacceptable—frozen, inflexible O-rings, or tiles repeatedly falling off—was common to both incidents.

This psychological mechanism is in play when we try to ferret out the waste all around us. There are things that are below our standards and that don't change for a long time. After a while, we stop seeing them. We don't see them! In fact, the lower standard job, housekeeping, or organization set-up, looks normal.

To see the anomaly with fresh eyes, we must do some exercises that will free us from that visual-mental mechanism. One of the best ways to temporarily cure the blindness is to visit other plants. If your company has multiple plants, arrange for maintenance professionals from one plant to be Lean consultants for another plant. This experiment is instructive, fun, and educational because, for a short time, when the wanderers return, they will be able to see the waste in their home plant more clearly.

Develop pointer lists

If all participants in a waste-elimination process are from one plant, an easy way to uncover the waste is to go through a structured process to strip away the blindness. To do that it is a good idea to develop some lists of pointers to possible waste. Development of the lists should be done in short meetings convened for that purpose.

Where and How to Look for Waste

Different meetings should have different audiences. But in all meetings, the audiences should be workers having direct contact with the process and its development. Supervisors, engineers, or properly trained workers, can facilitate the meetings. In some meetings (such as the first one on complaints) it would be best to have homogenous groups (maintenance in one meeting, and operations in another). For some lists it pays to have a diversity of workers from around the plant(s). For the second list (disruptive events), the best group would be a diverse collection of people from the area being studied.

The whole process is first to list specific things. In this work, generalities are not useful, specifics are needed. The next step is to pick items from the lists that appeal to everyone. The last step before action is to develop a project around waste reduction that is related to the list.

These lists are valuable. Even if you do not use each item to develop projects, the first go-rounds of the lists have value. And at the next go-round you'll not be starting from scratch. The existing lists are to be used as starting, or jumping off points. In a previously-quoted article by George Koenigsaecker, the point was made that the Leaning process was iterative.

Koenigsaecker said that each time a pass was made at something to make it Lean, new opportunities opened up. Once a certain level of leanness was achieved, the participants saw new opportunities that could not be seen from their old, fatter, vantage point. The Lean process needs you to work your lists a few times and then start over, because clearing out some of the fat will highlight new fatness that could not be seen before.

Pointer List 1: Complaints

One list is going to be of the major complaints. In this process, the ten top complaints will be listed. But what do people complain about? You know people complain but what (as specific as possible) do they complain about? Make a list of the things that most operations people complain about in relation to maintenance. At this point you just list the complaint and clarify it. There is no need to discuss it, defend against it, comment on it, or do anything else.

Chapter 24

You will find that these complaints have themes, so you can group them. One idea is to get a group of operators together, and talk about the different complaints. Use the group to develop the themes and populate them (with about ten complaints). Although any complaint can be used for this list, it might be helpful initially to stick to complaints about maintenance, about the machinery, or other related areas. The complaints might be about how maintenance is conducted, how the machines operate, failures that occur right after servicing, anything in this domain.

You could also work this list using maintenance professionals. You'll find a lot of the complaints are universal (but they may be stated in a way that's a little bit different). It's interesting (but not surprising) how you can be in widely different businesses and the complaints about maintenance will be the same. Both versions (maintenance, operations) are useful for the subsequent exercises.

Let's take the common operations complaint that maintenance folks are always driving around and not actually "doing" anything. That's a typical complaint. If we were going to address that gripe from the point of view of waste, what could we do? First we might try to quantify the problem (how much time) and do an analysis to identify the biggest contributors to the driving around. For fun, let's say we studied the issue and now know why people are driving around. Please notice that one thing we don't do is say that the complaint is not true. It does not matter in this exercise if the complaint is true or not.

One study found that the maintenance workers were driving around to get to assigned jobs and pick up parts. Different categories of parts are stored in different areas (there are separate places for things like conveyor belts, conduit, steel plate, structural steels, and small parts). The work is remote from the maintenance shop so that workers have to return to the shop for supplies, tools, drawings, instructions, and breaks.

To follow the process through to the end, as an example, we would guess, investigate, or brainstorm, solutions to the waste problem. Some people call problems, opportunities, and in this instance, that is exactly what they are. Some solutions might be:

• Visit the job, plan the materials, and deliver the parts to the job site before the job starts.

- Put more complete tool boxes and small parts kits on the trucks to reduce trips.
- Improve scheduling so that jobs that are near each other can be done in one trip. Pick all materials and tools together.
- Maybe use somebody else to get the parts, an operator, or an unskilled person.
- Build satellite store rooms with machine-specific parts, near where they are used.
- Build a single warehouse and tool room where everything is stored and locate it near where work is performed.

Any of these 'fixes' or projects could address this driving around complaint and the underlying waste. The one you choose would be thrashed out with people in the field and would be the solution that gave the biggest and quickest return for the investment, was least disruptive, and used common resources.

But what if your investigation showed that the driving around was due to federally-mandated inspections? You found that people were driving around doing visual inspections, and they were really working. The complaint is still valid as a pointer to waste. There is no blame here, just an observation of people driving around. Please note, it takes some character to hear complaints with this much detachment.

If the complaint is valid, consider whether any technology exists that would replace the visual inspections. Can instruments be mounted so that the inspection can be done remotely? Can the system be redesigned so that human inspections are not needed? Of course, any solution must comply with both the spirit and letter of the law.

If every repeated complaint were examined, ways might be found to uncover and then eliminate waste. Customer service might also be improved, but that is not the goal, it is just a by-product of waste reduction.

Let us now discuss complaints from a different point of view. Based on some distinctions from investigative work by Landmark Education it was proposed that complaints are misunderstood, and that most people think complaints are bad. But people complain because they have an underlying commitment to excellence, and the service that was provided to them, or the observations they made, did not measure up to that commitment. Somebody who complains about service in a restaurant has a model of service that the restaurant didn't match. So the fact that the service didn't match the model is not a problem for the complainer; it should be a problem for the service organization.

We can't make everybody happy—the goal is not happiness here, the goal is getting rid of waste. But we need to think about complaints in a slightly different way than maybe you do now, and realize that complainers always have something that they're committed to that's not being met, and that's why they're complaining. If they weren't committed to something, what difference would it make if they saw somebody driving around? What difference does it make if the machine is not fixed on the first go-round? What difference would it make if they weren't committed to putting out a high-quality product?

Pointer List 2: Disruption List

The next list is usually related to the first list and both may have common entries. Disruptions and complaints are usually related. Again, develop this list using operations people, and repeat the same exercise with maintenance people. Ask for a list of the ten most-disruptive events in operations. These disruptions can be pointers to areas where the equipment is less robust, or the process is not bullet-proof. Most of the items on the list might be beyond the scope of a Lean Maintenance initiative, and might be more suited for a production improvement project of some sort.

If maintenance is in good touch with what is going on in operations, then elements of the lists will be identical. The lists will diverge when operations discusses minor jam-ups, set-ups, and model/color changes, because maintenance generally is not involved in those disruptions. But from a waste point of view, these other areas might be ripe areas for projects.

Pointer List 3: What parts do you use the most?

The third list concerns parts with particular qualities of usage. Make a list of the parts of which you use the largest quantities. This list can be developed most easily by the storeroom. The selection is strictly the quantities of individual parts. List the specific parts (get all the way down to the part numbers). Choosing just 'bearings' is not specific enough. You must be specific with a part number or specific description. The list must contain details of the ten parts used in the largest numbers.

If you have a CMMS or stores system, ask for the parts with the greatest usage for maintenance purposes. If the data is available, ask that a column be added with data on where the part is used.

In a recent session on Lean Maintenance I overheard a comment within one of the teams. The participant said, "Wouldn't rubber work?" I asked them what kind of conversation would include the phrase "wouldn't rubber work?" The answer was, it was the beginning of a project conversation to eliminate waste. In other words, it's taking an item from the list and turning it into a potential project. Maybe the team could do an experiment and see if rubber would work in that application. If these ideas come up in the discussion, have the teams record them and move on. Do not let the team take time at this point to weigh the pros and cons of rubber. The goal is to get a list of ten parts in descending order of usage.

Converting the lists into projects is a later step. We need to start to look and see, which one of these complaints, disruptions, or parts usages, lends itself to a project (that would end up eliminating the underlying wastage). The question will be, why is something being used so much? Imagine how great a benefit it might be to answer that question. It would be a great boon for eliminating waste because, not only does it point toward eliminating the part waste, but it also eliminates the labor waste that goes into replacing the part.

There are two ways to approach the list-making step. At this stage you might not be able to answer some of these questions asked by these lists. One way is to just flat out guess what parts will be included on the list. Surprisingly, although inaccurate, your list will still yield a great group for the next step. The other approach is to find the right answers to these questions as rigorously as possible. That is a fine approach too.

Chapter 24

Pointer List 4: What are the parts you use that consume the most money?

To determine the parts to include on this list, have the warehouse generate a list of all the parts used for the last year or two, such that they are sorted in descending order of usage:

Calculation of period cost: Unit cost X Total usage for period

So, if the part costs:

$50 and the usage over 2 years is 31, then this part consumes $1550.

Cost and usages for all parts on the list are multiplied like this and sorted so that the largest numbers are at the top of the sorted list. The question really is "What parts are consuming the most money?"

This group is different from the first list. The first list was just parts used in large numbers. Some of them are only worth a dollar, so that you could use 1,000 of them a year and still not be in this group (parts that consume the most money). In another example, you might have one substation in the group that blew up last year and cost $500,000, that would make this group. In the middle of the list might be parts that are moderately expensive and have strong usage. These parts might include a $150 bearing, of which you're chomping through 175 a year.

A study done in the 1980s showed that 7.2% of the SKUs (stock keeping units—or individual part numbers) consumed 76% of the maintenance materials budget. Informal updates to this study show that the proportions have not changed. This result means that a very small number of parts consume all your money!

Think about the nature of the parts that are on this list. Usually these parts are either very expensive (with low usage) or they're moderately expensive but are used a lot. These parts are really good to use to uncover waste because they are the single largest consumers of your maintenance budget. If you mine gold in South Africa and you find an ore vein with ten grams per ton you are happy because that is decent quality ore. This list is like finding ore with 700 grams of gold in a ton.

Where and How to Look for Waste

What are some parts that are consuming a lot of money? Typical lists from maintenance professionals in different maintenance environments included:

- Consumables like fluorescent lamps in office buildings and hotels.
- Batteries, motor oil, and tires in truck fleets.
- Motor control components in factories (if sorting is to be done rigorously, we would need to identify the components by part number).
- Specific overhauls—A company had an overhaul example of an air cylinder that cost $7,500 and they were doing 50 or more overhauls a year.
- One power generator (electric utility) listed turbine blades.

The preferred approach is to go into the store system and run a calculation like the above, then sort in descending order, and then take the report, which will be a couple of inches thick, and keep the first three or four pages. I won't say any more on that instruction because it would be Leaner to write a query to print just the first few pages. If you did such an analysis every year, and looked into a few parts every week, then tried to fix the root cause of the usage, you could rerun the report every year, fix some, and catch some new ones. Some of the items on the list would not be good candidates, so they can just be skipped over.

Pointer List 5: What jobs are you doing that are repetitive?

What are the most repetitive jobs that the maintenance department does? What jobs does maintenance do over and over again? Create a short list of the top ten. jobs on this list. They do not have to be expensive, just repetitive.

Some examples of jobs done over and over might include:

- Changing lamps.
- Responding to heat and cold calls in a building.
- Repetitive changing of both the rolls and the castors in conveyers.

- Chain lubrication.
- Sharpening blades.
- Maintenance at certain intervals, so include oil changes and even many PMs. This aspect may sound sacrilegious, but we know that the idea that 70% of the PM that we do is wasted was discussed in the chapter on PM. There is no intention to discredit PM in any way, but we want to be smart about this because there is waste here.
- Unclogging pipes.

Pointer List 6: What jobs consume the most labor hours?

Pick the jobs with the most labor hours, including repair jobs (unscheduled shutdowns) and scheduled overhauls. Large jobs might also be contracted out, so that the department can keep up with its day-to-day service commitments. Be sure to include jobs done by contractors.

In sessions on Lean Maintenance, certain jobs seem to be large jobs for everyone.
- Inspections (added together—but we want to separate this work into its constituents).
- Catalyst changes in a oil refinery.
- Shutdowns (both scheduled and unscheduled).
- Installations.
- Lining mills (in an ore processing center).
- Equipment overhaul.
- Pump rebuilds.
- PM.
- Meetings. Consuming most labor hours of all in some plants. It is important to improve your meeting expertise. The rule that we discovered (and there is some research on this subject), is that the more comfortable the room, the longer the meeting for the same number of agenda items. So, if you want a longer meeting, serve coffee, if you want it even longer, serve buns or donuts. But if the goal is to cover an agenda in a given period of time, you want to be conscious of the fact that if you take the chairs out of the room, it will make the meeting shorter. If you

have the meeting in a room that's 30 degrees either F° or C°, it generally is a shorter meeting for the same number of agenda items.

Pointer List 7: What are the jobs that cost the most, that are done in a given year?

This list is frequently very similar to the previous list because the most expensive jobs usually require a lot of labor. This list will pick up jobs where the labor content is low but the parts are expensive.

Some jobs that fit into this category include:

- Turbine rebuilds.
- Major pump work.
- Turnarounds, shutdowns, and outages, lead this list. Most companies that have big operations have to shut down to do most maintenance work.
- Major jobs inside shutdowns. You might drop a level for this exercise and look into the shutdown, and see what are the jobs that are really costly?
- Software. This cost is an unusual choice that came up on a list in a Lean session. Participants saw the charges for buying and training, and the annual maintenance fee, and felt that software was fat for their business. The other question with software is, how do you know if you're receiving a benefit? How can you measure the return on investment? It's kind of an interesting way to think about as a waste thing, something that requires a license is a waste, if you don't use it. And it's worth investigating whether there's some other way to accomplish the same ends or not.

Particular subsets of the possibilities of waste are embodied in the answers to these questions. The approach is a kind of a microscope; and it's also a way to look at something that is different from your normal way of looking. The ultimate goal is to enable you to develop a lean eye. A Lean eye is an eye that sees if things are optimized. That's the goal of this discussion, that you start to see what is possible in optimization.

Chapter 24

One example of a waste list in action: Simplified Compressed-air survey

One complaint (from list number 1) was that the compressed-air supply was not consistent (pressure varied with load) and that inconsistency resulted in uneven quality in the parts that were produced. The air supply varied in pressure and quantity. An investigation showed that there were numerous minor leaks, but no major causes that could be identified. Before going through the process of developing specifications for a new compressor, the team decided to do a compressed-air survey and leak reduction effort. They wanted to see if they could solve the problem without buying another compressor.

Members of the group read an article by Paul Studebaker (a noted expert in utility reduction), and decided to follow its recommendations. According to the article in Plant Services, titled *Compressed Air Isn't Free,* by Paul Studebaker, statistics from the Compressed Air Challenge (CAC) and DOE (US Department of Energy) were confirmed by compressed-air system experts, saying that the average facility has 30% to 35% compressed-air leakage if it hasn't taken any recent action. And a survey by the Office of Industrial Technologies says 57% of facilities have taken no action on compressed-air problems during the past two years.

The article said that savings from identifying and fixing even a few leaks is great, and the ROI (return on investment) is off the charts. Usually, the only investment is a few hours of labor, and possibly some fittings, and good-quality pipe dope. The same article pointed out that good-quality pipe compound is superior to Teflon tape because it fills the voids better.

The compressed-air survey is one of the simplest projects to do, and it gives almost immediate payback. The payback is in 2 areas. One area is in the immediate savings of electricity. At 0.75 KW per horsepower, saving even small amounts of compressed air will produce savings.

The other savings is in compressor capacity, which was the goal for the team, to eliminate the need for another compressor. This savings is not immediate, but can reduce the number of compressors needed to service a plant. In other instances, the savings in air load is enough to cancel plans to add a compressor. Sometimes there are operational benefits to the quality and consistency of the compressed air because the system is less strained. Another great benefit is where a compressor was running almost full time and can now be used for its original purpose as a back-up or alternative source.

The article pointed out two major ways to conduct the survey. The low-tech way is to have a crew come in on a Sunday (if the plant is not running) or in the evening (if the plant works only day shifts), and just listen for leaks when the rest of the plant is quiet. Soap and water can be used to detect the leak positions. The crew should hang tags and tighten, repair, and remake the leaking joints. Of course, the rule is that pipes rarely leak from the middle!

The high-tech way is to use ultrasound. Even pin-hole sized leaks whistle in the ultrasound frequency range. Ultrasound is highly directional, so that the microphone or detector will point directly at the leak. Ultrasound would be appropriate where the plant doesn't shut down. And ambient noise is less of a problem with ultrasound.

The team wanted to get started immediately so they decided to go the Sunday route. Teams went around one Sunday and tagged any leaks they could hear. On the next weekend they turned off the air system and repaired the leaks. The benefit was immediate and obvious. The currently-installed compressors now could easily handle the load and supply the needed volume and pressure. In fact, the old back-up compressor now only ran infrequently (whereas it had been running full time before the project).

The crew returned a month later with an ultrasonic gun and found another large number of even smaller leaks. Needless to say, the customer was satisfied and the problem is now in the past. The team did not record electricity savings, although those also were substantial.

CHAPTER

25

Lean Utilities

Many articles have appeared in the press about the problems associated with global warming, and looking at the carbon footprint of various business alternatives. This issue will not go away. And every enterprise has to address it. In fact, industry will either be forced or will, by itself have to develop, sustainable models of manufacturing. One of the intentions of this book is to support organizations in that effort.

Although all the maintenance departments in the world, even acting together, cannot solve the global warming dilemma, there are things that each of our firms can do to help, right now. Many of the Lean projects use the returns from improved use of energy to fund themselves.

By definition, Lean maintenance is also energy efficient. Energy is one of the inputs into almost any manufacturing process. Wasteful energy practices are both fat and, in today's public climate conversations, almost antisocial. The body of knowledge in fighting global warming is growing rapidly so there are always ideas out there worth looking at.

In the previous chapter, in the discussion on reducing compressed air leaks, the driving force was capacity and not efficiency. Over the long haul, the energy savings would probably dwarf the savings from an avoided compressor installation. The other advantage is that the energy savings would be there every month, after the leaks were fixed. In good times and bad there would be that small savings. Utility savings are like that. They are small rivulets of water flowing toward the bottom line. As you complete more and more projects in this area, the rivulets join together and form creeks, then rivers of savings.

Lean Utilities

One of the experts in the field is Eaton Haughton of Econergy Engineering Services Ltd. who helps owners of buildings such as hotels and offices, save money. He estimates that utility savings can be obtained from the following activities:

- Monthly Utility Bills Analysis (10%)
- Improved Corrective and Preventive Maintenance (15%)
- Improvements in Staff (and customer) Awareness (10%)
- Reduced Outside Air Infiltration to Conditioned Space (10%)
- Clean Heat Transfer Surfaces (15–20%)
- Retrofitted Fluorescent Lamps with electronic ballasts and high-efficiency tubes (40%)
- Changes from CFC, HCFC and HFC refrigerants to natural refrigerants (15%–25%)
- Power factor improvements (20%)

One of the largest electrical loads in an office building is lighting. It is also an area where (depending on what kind of lighting is currently used). Savings can be immediate and relatively simple to achieve. In looking around at the various parts of your buildings, it is easy to see areas that are used only infrequently (consider occupancy sensors). Are there areas where changes to more efficient lamps could be made easily? Several ideas are no-brainers. If you have any incandescent lamps left (at home or in the office), replace them with compact fluorescents.

Ideas for projects in Lighting

- Time switches (10%)
- Occupancy sensor switches (30%)
- Photo-electric switches (30%)
- Electronic ballasts (30%)
- High-efficiency fluorescent tubes (20%)
- Retrofit with fluorescent lamp reflectors (15%)
- Compact fluorescent lamps (60%)

The best thing about making lighting more efficient is that it reduces the air conditioning load at the same time, so you have a double bonus. Of course, in a cold climate it increases the heating load but lighting is still an inefficient way to heat. While you are looking at lighting, make sure no one is still using old CRT computer screens. Flat panel displays use less energy, are easier on the eyes, and will reduce the air conditioning load.

Lighting survey

One of the simplest Lean projects is to do an inventory of the lighting in the plant and office areas. You will immediately find opportunities, unless someone has beaten you to it!

One of Eaton Haughton's projects in an office building was to replace four Magnetic ballasts with (1or 2) Electronic Ballasts

This project cost: $125, including the costs of 3 X 36-W T8 tubes, electronic ballasts and reflectors

Annual savings were US$140/yr/fixture.

Lighting demand was reduced by 40%–60%.

Air conditioning load was also reduced.

Insulation

How many times have you thought it would be a good idea to add some insulation to your home? Insulation is also a good idea in factories, office buildings, and other places of business, including insulating of process heat sources. Many factories have expansive roofs that pick up significant heat loads. Solar collectors for both electricity and for hot water can reduce some of that load. Insulation can help with the heat load that is left. Some of the advantages are:
- Substantial reductions in roof heat load
- Reduced roof maintenance $$
- Improved equipment life
- Costs of only US$2.00/square foot
- Savings of 20%–30% on building cooling costs
- Payback = 4–5 years on energy cost reductions

Solar energy is a hot area for development in sunny areas of the world, and solar heating can often be used to supplement process heating. Payback can be swift, especially if the hot water is needed remotely, in relatively small amounts. An overall energy reduction strategy can be good for the company in more ways than just increasing profit. Even small projects can be publicized, and can gain good local press for the organization. Some governments also offer incentives and tax credits, which make some marginal projects economically viable.

Alternative Energy Source: Electric to Solar refit for domestic hot water.

One 3m^2 (32ft^2) solar collector can deliver up to 300 liters (80 gallons) of 50°C (120°F) hot water per day (in a sunny area of the world).
- Equipment can be retrofitted to existing electric or gas water heaters
- Saves energy and maintenance dollars
- Cost depends on system size (approximately US$1,000/hotel room for example)
- Payback = 12–18 months

In another example, you can get free hot water from the heat exchange in air conditioning or process cooling units. In some installations you need to look at the whole process and use the wasted energy in other areas. Co-generators, where you use the heat from an electric generator, are an excellent example, in which the extra return from the steam or hot water might make the whole project feasible.

Producing free hot water with an air conditioning heat exchanger

- Delivers 10–15 gallons (130 F) of hot water per hour, per ton of air conditioning.
- Saves water heating energy dollars.
- Saves about 8% of AC electricity, due to a more-efficient cooling effect with water vs. air.
- Capable of meeting the total hot water needs (of, say, a hotel).

Chapter 25

- Cost of approximately US$150 to $500 per ton, installed.
- Payback = 12–24 months.

Thermal Energy Storage

Thermal storage allows cooling to be done at night for use the following day. This arrangement improves efficiency for utilities because it allows them to provide more energy with the same generating capacity. The resulting lower cost is passed on to customers in the form of lower rates.

The SABIC headquarters in Riyadh, Saudi Arabia had a significant, below ground, ice making facility that was used to minimize the load during the hottest days and as a back-up system to the chiller units. The headquarters housed the data center, so it was important that it never lost cooling in the hot, Saudi Arabian, climate. Making ice provided the back-up, and lowered the energy cost for the whole facility.

Thermal Storage Saved Money in Several Ways.

- By decreasing or eliminating chiller operation during Utility Peak Periods.
- By displacing energy use from the expensive peak, to low-cost, off-peak periods.
- Equipment capital cost can be reduced, particularly for new installations.

Fun thought problem: Take a look at the parts that were replaced during a breakdown. How many KWH went into making the part that was just thrown away? Try to visualize the process back to the beginning (iron ore, petroleum), and through all the steps to the finished part. Even a small part has quite a large energy tail.

CHAPTER

26

Translating Lists into Projects

After conducting the meetings and developing the lists, the question is, how do you turn the lists into Lean projects that will pay dividends? As you lay out the lists, some of the entries will recommend projects directly, some will require additional research, and some will just not stimulate any thoughts or ideas.

The first projects to tackle are the ones that recommend themselves. They might even jump off the page at you. On the first go-round, don't concern yourself with the difficult ones, or the ones that require cultural changes, excessive time, money, or exotic resources.

One thing that is important when looking over the lists is specificity. To do the analysis in this stage we must isolate the source of the demand for the resource. Money is spent down in the trenches of the facility—if we want to reduce that spending, we have to get down into the trenches with it. This need for specificity is a major reason why this approach to Lean Maintenance is best carried out by the workers already doing their jobs in the trenches.

For example: if the list says only 'Motor control components', we don't have the information for analysis because the parts usage could be coming from a variety of demands. This analysis is easiest if you can get down to one demand. In some areas, the first step is to do the research to make the lists more specific.

Assuming the lists already are pretty specific, pay particular attention to the items that pop up on several lists. These concerns are usually important, expensive, or particularly annoying.

Examples of the leap from a list item to a project

Source list Complaint: Too cold in room

Chapter 26

Possible projects:
- Test replacing windows with energy-efficient ones.
- Insulate areas of heat loss.
- Plot sun movements and determine height in sky. Change geometry of awning so that winter sun can get into room.
- Photograph the room from inside and outside with infrared camera, and caulk the places where heat or air are escaping.
- Move desks around to get people out of drafts.
- Try putting streamers on diffuser to visually show when it's on, or placing a big thermometer on an interior wall away from drafts.
- Issue sweaters or blankets ☺

Measures: Before and after temperature readings, count the number of complaints from users. Measure the KWH consumed by area heaters before and after changes.

The question is: how do you evaluate the potential projects and chose the best one? Look over the list and decide which projects have the highest probability of success. Estimate which potential projects are the quickest to do, will show results soonest, easiest, least disruptive, or fastest pay back? Another approach is to decide which project can be done this afternoon? Of course, look at the consequences, or down sides. What is the down side?

In this research, streamers might be the quickest, or taking an infrared picture (assuming you have access to a camera). Clearly, the most expensive approach would be to replace the windows or insulate the walls. In between these ideas would be moving the desks (depending on the cabling). But moving the desks would be disruptive. Using an infrared camera would have a higher probability of success than just adding streamers, or a big thermometer on the wall. Each project would have pros and cons that the team would discuss and weigh.

Source: List of Parts usage: High pump usage (high-cost job, frequent job)

Possible projects:
- Try a cheap pump in the application so that each pump change is lower in cost.

Translating Lists into Projects

- Try a better pump that you think will last longer.
- Design a PM program to extend pump life and PdM to alert you when a failure is imminent.
- Add a back-up pump loop for quick switch-overs and easy service access.
- Disassemble and study the pumps that fail and determine failure modes.
- Call in a pump distributor or manufacturer to look at the application and to make recommendations.

Measures: Cost per year, downtime costs (both before and after changes), energy costs, number of service calls.

In this example, the easiest alternative, with the least down side, would be to get a vendor to come in and look at the application. Most companies would also try their hand at a PM/PdM system, with or without the advice of the OEM or dealer. Probably the next-best alternative would be to take the bad parts to an expert and get an opinion about how and why the pump failed. Putting a back up loop in would usually be the most expensive. Testing a cheap pump or a better (more expensive pump) might be pretty easy, depending on the criticality of the application.

Source: List of High-cost jobs—Follow the money: Motor Oil is one of the largest consumers of dollars in a bus fleet (high-cost part)

Four possible projects:

- Investigate different ways of buying oil (bulk, versus drum, versus gallons),
- Try oil analysis to extend oil-change intervals,
- Look for better grades of oil that last longer,
- Investigate piping oil directly to PM bays to eliminate waste, contamination, and ends (unused oil left in drums).

Measure: quarts of oil used per 1000 miles, cost of oil per 1000 miles, engine repair costs.

Oil analysis would offer immediate feedback, and is both not disruptive and inexpensive. Analysis also has other advantages. Improved oil is another, albeit more-expensive choice. Changing the way you buy the oil is the next choice because there will be clear savings, and if you can go to a larger container without problems, you're also throwing away less packaging. Piping the shop is the most expensive and disruptive, but also offers the greatest savings in both labor, cost per gallon and reduced contamination.

Source: List of Repetitive jobs: grinder going out of alignment
Three possible projects:
- Disassemble grinder head and look for excessive wear or play. Repair or replace worn parts.
- Ask the OEM for their opinion or their help.
- Look at jobs that are running to see if there is a relationship between the job and the problem.
- Develop a procedure for alignment and set-up, and certify that everyone who touches the grinder is explicitly checked out and follows standard procedures.

Measures: Call backs, quality ratio
The third of the above items is the least disruptive and requires the least outside resources. It might also have the lowest probability of success. The first item is the second easiest but would require spending money on parts. You could do both items one and two, and you could also contact the OEM and have their experts look at the worn parts and offer opinions. The last would be the most expensive and disruptive, but might also have a high probability of success.

Let us return to one item from the list developed at Alcoa:
- Time spent going to jobs and arriving unprepared. The maintenance people frequently don't have the materials or tools, and then have to leave the job to collect the stuff.

This item is more complicated than the pump problem, and there is a wide range of possible responses. One approach is to develop a project to work-sample (take random snapshots of what is going on) some jobs to assess the extent of the problem and to help quantify the amount of time lost. You could ask the 'old-timers' what they have done about this problem. You could find out whether the problem is universal.

Translating Lists into Projects

Another alternative is to take a random sample of work orders and compare the Bill of Material and the tool list to what was actually used on the job. The solution to this kind of problem could be simple (like reconfiguring the work order to print the bill of material for the tradesperson or complex (like instituting planning from scratch).

The point is that it is easy to structure a short information-gathering project. Once information is gathered, you could meet and examine solutions based on the facts that you have gathered on the ground. The first step in most projects (except those with the lowest hanging fruit) is to gain more specific knowledge of what is happening.

The big challenge is to translate these lists into potential projects. This ability can help make the leap from the complaint list, for example, to an actionable project that studies the complaint, fixes the complaint, or in some concrete way, impacts the complaint.

When you start to focus on one item from a list, consider these questions to help gather needed information, but only the information that applies to the problem:

Economics:
- What is the cost of the old way of doing business?
- What is the cost of the down time? (Total Down Hrs x Rate)
- What is the total cost per year?
- What is the return on investment of a projected improvement?
- How much can we/(should we) spend to fix this problem?
- What is our investment in this asset or process?

If the issue your team wants to look into is a reoccurring breakdown, the following questions might aid your team discussions.
- How disruptive is this breakdown?
- Is this breakdown caused by an action of a maintenance person (iatrogenic)?
- If there is a TPM program, was this a failure of the TLC (tighten, lube, clean program)?
- Is this breakdown caused by an action of an operator?
- Does this breakdown cause mechanical or electrical problems elsewhere?

- What is the honest opinion of the maintenance old-timer' experts.
- Is the root cause a faulty or inadequate maintenance procedure?
- Is the root cause related to inadequate training in any maintenance skill?
- How often does the incident occur?
- Is there a pattern or trend?
- What is the mean-time-to-repair (MTTR).
- Compare down hours to MTTR. Is the response fast enough?
- What is the mean-time-between-failures (MTBF).
- What was the mode of the failure?
- Is a structural analysis of the broken parts indicated?
- What happened just before the failure?
- Why did the breakdown take place?
- Was there a failure of the PM system?
- Do the PM task lists look at this failure mode?
- Are we looking at the root cause or a symptom?
- Is the root cause related to an inadequate part specification?

Some of these are tough questions to answer. The thing to remember is that, whatever you find, it is not personal. For example, if the root cause is operator error, the question is not who, but how and why. We are looking at the system that allowed the failure. The focus is on eliminating the failure mode.

Conversion of items from the lists to possible projects

Review all your lists and choose one item that looks as if it is solvable, or where the potential return is quick. Describe some potential projects, and focus on the few with the highest probability of success. Use the sheet to name the projects, and review each project based on:
Probability: What is your best guess of the probability of success.

Low Cost: Low cost is just how much money is needed. The lower the better.

Translating Lists into Projects

Short Payback: Short Payback, or how long will it take to pay back the investment. Shorter is better.

Tools and Materials: Tools and materials available and that nothing is needed that is not in stock or easily obtained.

Return on Investment ROI, which is a percentage of the returns. Higher is better.

Risk: Low downside risk is preferred, including disruptions to process, hazards and safety, and environmental impact.

- Some of the best projects are so simple they are called 'no-brainers' or 'low-hanging fruit'

Use a form like that below to check out the items on the lists. You would take one item at a time, that seems to recommend projects, and work it through. Always look for the low-hanging fruit. Save the work-sheets because, at a later time, the other projects (skipped over in the first look) might become more appealing.

This is a two-pass process

On the first pass: Use brainstorming techniques to make a list of potential projects for one item from one of the lists. Do not discuss the items at this time (otherwise you'll get bogged down and not generate enough ideas). If you are not familiar with brainstorming look up the simple rules on the Internet or get a book and share the rules with the team.

On the second pass (can be a subsequent meeting): Take some time to discuss each project in relation to the five criteria mentioned. Weight the resources needed to do the project and the returns you accept. It is essential in the beginning that you chose projects with a high probability of success. Pick the one project that your team agrees on, and then proceed to the next step.

Chapter 26

Project Brainstorming sheet

Problem	List Name:	Date:	LOW $	Short Pay	T&M	PROB.	Risk
	Item from List:						

Choosing the projects to take to the next level

The final step after the second pass is to choose one project to look into, plan, engineer, and finally (when you agree it meets the five criteria) decide to execute. *Beware of over-analysis.*

The first key is **"do no harm."** So eliminate all projects that have excessive risk, no matter what the return potential. Sometimes the risk is not obvious. Take the time to make sure the risk is small. Risk occurs in a variety of ways, including disruption to the process, impact on quality, or production rate. A large part of risk is hazards. Look at both the safety and the environmental effects.

The second key, which is especially important in the beginning stages of the Lean Program, is to choose projects for which the probability of success is high (PROB. above). It is best to get ten or more successful projects before the program is discussed publicly outside maintenance, in any detail. Before any publicity, successes are essential.

After the first two keys are satisfied, look at quick payback and availability of resources. A quick project is great but not if it needs expensive outside resources. Save those projects until the whole program is more mature. If a choice has to be made, go for the project that can be done with existing resources, even if the payback is a little slower.

CHAPTER

27

Unintended Consequences

There is a dark side to fixing problems. Sometimes the cure is worse than the disease. In our world, that is when the Lean project produces results that were not intended in the design of the project. These outcomes can contribute more fat than the project was supposed to reduce in the first place. Thinking about what can happen should be part of every maintenance project, or even repair, and certainly behind every possible Lean project.

In a recent Maintenance Management class, the team was trying to eliminate the complaint that the drivers of the dump trucks in a water utility sometimes took home the keys to their trucks by accident. All kinds of effort went into trying to manage key locations, which was a minor but frequent problem. The fleet group decided to attach a large aluminum plate to each key ring (similar to what is often done with rest-room keys in gas stations). This idea was a wild success, and was soon adopted by other departments.

However, a few months later the fleet department was screaming because they had to replace large numbers of ignition switches. The switches were not designed for the extra weight of the plates. The solution to one problem thus created unintended consequences. The ultimate solution was to use a thinner gauge of aluminum. Although the consequence was minor, it does demonstrate how solutions can cause unintended problems and get out of hand. The consequence can occur in someone else's department (making it hard to track and manage).

Change always has both good and bad results, even when the change is supposed to be an improvement. There is a rule of unintended consequences that says you can't always see the consequences

Chapter 27

of your actions. The solution for today's problem might be tomorrow's nightmare.

For example, let's say we decide to reduce the maintenance cost on a boiler. Months later, the heat exchanger starts to degenerate, and then we get higher utility costs as a result of the lean maintenance decision. Such cost increases often happen when people cut costs in maintenance. Management says, cut the maintenance budget by 10 per cent. The consequence is that downtime starts to creep up by a little bit, because maintenance wasn't there when they were supposed to be, or didn't have the PM hours they were supposed to have. The effect does not follow directly (in time), and can be subtle. To guard against unintended consequences you must look at everything that's connected to the chain that you're working on. Some of the results are going to be unintended, when you didn't even think of them as being related.

The conversation about unintended consequences reminds me of an arcade game at a kid's restaurant called Chuck E. Cheese. In that game there's a field of holes and there's a mechanical gopher that sticks his head up. Your job is to whack the gopher on the head with a stick or a mallet, and you get a point for each hit. As the game goes on, the gopher starts to move faster and faster, and eventually two gophers stick their heads up. You have to whack both of them, and you can do that if you're really, really, fast. That's the gopher game, but it also could be called the game of making maintenance management Lean, because you do a project and make a cut in one area of what looks like waste, and then another gopher sticks its head up in another area and costs increase.

Here is a list of questions designed to uncover unintended consequences and hidden relationships between things, processes, and procedures. Before undertaking any except the simplest project, have a round table discussion between people from maintenance, engineering, and operations, and bring up some of the relevant issues from the list below (of course don't be limited by this list):

• How will the project impact operations?
• What could go wrong?

Unintended Consequences

- Does the project force the failure or accelerate the deterioration of other parts of the process?
- Does the project affect the life of any component in any part of the process?
- If we did have a problem, can it be bypassed with a back-up or standby unit?
- Is there a scrap or start-up exposure?
- Could this project cause a failure of the operations system?
- Are operator training or knowledge changes needed?
- What is the impact of this project on the customer (internal/external)?
- Can we afford to have this type of project go wrong?
- Can this event have any effect on product quality?
- Will this project have any impact on employee morale (maintenance operations, other)?
- Is there an impact outside the maintenance sphere of influence (environmental, competition)?
- How high should the priority be to proceed with this project?
- Is there a regulatory or legal impact to this project?
- Is another logical business decision indicated (outsourcing the part, sell off product)?

Risk evaluation is a team exercise. One of your insurance policies against doing something really stupid is to have some people from other parts of the organization, who know about the process from different angles, participate in the Lean discussions. These meetings are good activities, with people that are not even involved in maintenance. Those people are not as blind as we are about certain things. They may ask stupid questions, but sometimes the stupid questions are really the ones that should be asked. We're so sophisticated about maintenance issues that we don't ask the stupid questions ourselves. Sometimes we ask the sophisticated questions, the answers to which are irrelevant, but not the stupid ones like, "Why do we do that?"

I was cooking a roast and I cut off the end and put it on top of the roast. And my former wife said, "Why are you doing that?" I said, "I don't know, that's what my mother did." So I called my mother

Chapter 27

and I said, "Mom, why do you cut off the end of the roast and put it on top?" And she said, "I don't know, my mother taught me that." So we called the nursing home and talked to my grandmother and asked, "Why did you cut off the back of the roast and put it on top?" And she said, "Oh, because my oven was so narrow."

So at least 3 generations of people cut off the back of the roast and put it on the top, and that's how things sometimes develop and how they stay in place. And no one was there to remember why this or that activity was done. But sometimes making the process Leaner has unintended consequences.

One of the worst industrial accidents ever, happened in Bhopal, India, because people didn't understand some of the safety systems in that plant. The systems were taken off line for service, and there was no urgency to put them back on line. There were quite a few causes for this particular disaster, and one of them was a deluge system that was designed to cool down the methyl isocyanate (MIC). This system was taken off line because the people didn't know why they needed so much pumping capacity. They knew that they didn't ever use that amount, and they didn't understand what it was for.

Obviously, you're not going to do something in that domain but you need to understand that although you don't know why they're there, it's vital that they be there and working. And then there are other systems that ought not to be there, and you've got to have someone on the team who can distinguish between the two.

Lean Projects: Organizational Impact

One of the jobs involved in having a really Lean Maintenance program is preparing the ground for change. Like any transplant, the business system will reject any program for which there is no structural change in the business processes. Lean is no different. Of course a manager, or even a supervisor, can initiate Lean Projects within their areas without much upper management support (or knowledge for that matter), because the managers themselves become the structure to keep things going. But a Lean project does not a Lean program make.

Did you ever notice how many efforts are personally driven by an individual? When that person leaves, the program often dies out. But if the management structures (things like incentives, reports, meeting agendas, speeches, etc) of the company are changed, and the change is bedded down, the champion can leave and the program can stay in place.

Getting your ducks in line is important. One of the ducks is, what are you trying to accomplish? Are you just doing some projects for fun and departmental gain, or would you like this effort to morph into a full-blown corporate initiative? Either is fine, so long as you know what you are doing.

Many firms have systems whereby improvement projects are assigned to coaches or champions. Can you have champions or coaches, given your business and attitudes at home? Coaches are senior people who have agreed to be sounding boards, devil's advocates, teachers, and obtainers of resources. Having some coaches supporting the Lean teams increases the probability of success.

Chapter 28

In business you learn to realize that you can't change what you can change. One of the things about maintenance people is that they tend to be secret, or closet, perfectionists. They would like something to be perfect but they've been in the field too long. The result is a tendency to complain about stuff that they can't change, and not change the stuff that they can change. This is another version of the normalization of deviance discussion we had about seeing waste.

What about changing a management structure that you see as wasteful? You might know from experience that some things, you can't change. Some of the systems and procedures you don't have a shot at changing, so let's not be preoccupied with that stuff. Instead, let's be preoccupied with the stuff that's under our control. Then we can do projects, follow the rules, and publish the results. If you follow that route, people know that you're on the Lean bandwagon and are being successful, and then the next time you ask, it'll be a little easier to get cooperation.

Lean Maintenance is more of a tactical approach than a strategic approach to managing maintenance. I don't want to give the impression that big things should not be taken on. It's just that Lean Maintenance team meetings are not the arena for those discussions.

If you are ever involved behind the scenes at a charity, anything from an expansion campaign at your local church to a larger charity, you'll see a particular approach to raising money. This approach can be a useful model for us when we start to talk about Lean Maintenance initiatives. When a charity starts a big campaign to raise money, they have what's called the silent part of the campaign. During that period, they raise money from their big donors. When the charity rolls out the fund-raising drive for everyone else, maybe 60 to 70% of the money has already been raised.

Everyone always wants to see that thermometer (that displays the status of the fund raising) getting near the top. People will say "Oh, but if we're that near the top, maybe my $100 will make a difference." The charity organizers also publish the successes, "We're doing this, and we're doing that. People are lining up to give us money." They say. And it's the same in the maintenance world; we want to look like winners. So, before anybody even knows about it, we've done ten lean projects in our department and we've gotten results that we can show, document, etcetera.

If you choose to take on Lean maintenance formally or plan to go formal after a few successes, then there are three documents for a Lean Project. One is the proposal, the second is the raw data, diary, and real-time narrative, and the last is the results report. The proposal will transform into the final report (at least as far as sections, and some of the starting information). The raw data and narrative will feed the final report with facts and stories to make it interesting, readable, and believable.

To write a proposal to do something, you have to start with a name, and this is very important. You must pick a really good name for the project. A good name is one that speaks to the customer. If you pick a boring name, who's going to want to do it? The cool name is what will get people enthusiastic.

If you were in a position to approve urban construction projects, which one would you rather get involved in; The cross-town tunnel or the Big Dig? In our world, would you rather fund the PM improvement Project or DEEP—Downtime Elimination Education Program or PIE—Profit Improvement Effort? (With PIE you have the added advantage of a good metaphor "We're going to improve your slice of the pie." And when you give your presentation you can bring in the pie and cut it up.)

In the next chapter we'll discuss the need to be clear, when you design your project, about what kind of waste you are going after? How will the improvement be measured? Remember, the waste will have to be measured before the project and afterward. So, we have to be able to do both. What will the situation look like after it's done? Can we describe what it'll look like? It'll be running smoother, it'll be doing this, and it'll be doing that. After that section, translate the impacts into money. What's the cost and what are the benefits? Finally, be sure to include a cast of characters.

Human relationships

There is one big issue as you get deeper into the organizational waste of your business process. The issue concerns changes that

Chapter 28

impact people's turf. The consultants who do business process re-engineering deal with this aspect all the time. They work to re-engineer some stupid processes out there, where they just kind of grew up, and then encounter all these different reporting needs here and there, and very, very strange kinds of relationships, of who can do what, and who can't do what. They recommend major shifts in job roles and responsibilities. The issue they face is that the existing system has a constituency because it gives people power, and people are used to it. The re-engineering might change the power, or even take it away.

The consultant finishes the business process review and presents it with the returns on investment and head count reduction. The gap is what is presented to management, who make whatever decision is made in the service of the greater good to the whole organization. The key is that this kind of process is always authorized at the very top of the organization. If our Lean initiatives take us into that realm we may have a problem.

To have the influence needed for a procedural change you have to go pretty high in the organization. But the politics are a little sticky for people inside. An outsider (like a consultant) can have those tough conversations; if they get into trouble, they just don't get hired again. That's why outsiders are brought in to have those conversations because they can. As an insider you have a problem. Some insiders could ruin their relationship with a person they've known for 20 years. And it's impossible to minimize the negative consequences of that result.

Those personal relationships are valuable. In addition, you might need something from that person tomorrow. And that person might need something from you the next day, so you always need to keep those relationships good.

But if you are going to initiate a change, you need to initiate a manageable change within prescribed limits and get a success. And that's what people will start to learn from the stories of success.

CHAPTER

29

Once a Lean Project has been Chosen for Refinement

Now the fun begins. You have made a list of possible projects. You have evaluated each potential project for the six areas: high probability of success, low cost, short payback of investment, tools, materials, and whatever other resources are available, good ROI, and low risk.

In this refinement stage, if a formal project is to be proposed, we have to write up the project proposal and prepare a verbal presentation to sell the project to management (if that is necessary). In the first few projects management permission is generally not necessary. Any proposal write-ups would be for practice. If you have a champion, let them be the interrogator to help you practice for the real management presentations given at later stages for larger Lean projects. By the same token write up the successes (and failures) for practice and for your Lean Maintenance archives.

Not until after that will we get approval of the proposal. Then we need to plan the project (similar to the way that any maintenance job would be planned). We can then execute the project and collect the data. The final step is to study the results and prepare a Project Report with the facts, the figures, and the story.

Measurement of the project

A good Lean project will have a before and an after picture (verbal, numerical, or even a picture). It wouldn't be very convincing if a weight-loss commercial showed only an 'after' picture and someone saying they were really fat before using the product. Lean Projects need baseline measurements, such as what was the reading before any change was initiated? What was the reading after the changes? What does the difference in the readings mean? If we observe that a process draws 550 KW before the project and 435 KW after the project we can see that we have saved 115 KW. Then we would look at utility rates and give a best guess or an accurate calculation to quantify the savings.

It is important to choose a way to measure the success or failure of any project. Sometimes, the most difficult part of a maintenance project is determining how to measure it. Even if it is difficult to measure, it is essential to develop ways to measure every project. Your cost analysis will be based on the measurements that you choose. For example, to get the best data, if your project involves energy savings you may have to mount a watt-hour meter to measure electricity usage before the project starts, so that you have a solid baseline.

The measurements might prove that the project is a bust. What do you do? Before you do anything, teach the team the importance of negative results. As a result of this project, you now know something you didn't know before. Work to create an environment where negative results are "OK," and no one gets sanctioned for picking a project that doesn't work out. Always cover negatives in your team meeting with a positive attitude. These negative results are so important that you should recognize them as fully as you recognize positive results.

The Lean coach

A management coach (also known as a champion) will: provide money that was budgeted, run interference, get supplies and help, provide faith during the dark periods, arrange access to assets and resources, and do whatever else is needed. Doing the project however, is not the coaches' job. Are you using coaches? Coaches increase the success rate of Lean projects. If the Lean Project is being carried out by workers as it is designed, then someone who knows the ins and outs of the procurement system, and production schedule can sometimes offer a deal-making advantage. Having a coach also gets other people involved in the Lean process. The coach does not have to come from maintenance but should know the ins and outs of getting things done in your company.

Cost benefit analysis

Lean Maintenance strives to save as much real money as possible, consistent with good, long-term maintenance practices and safety. In performing a cost analysis of the project, your costing should be based as much as possible on measurements and as little as possible on guesswork. In the proposal, be careful to make an accurate promise about savings. If anything, be conservative. Fully worked-out economics are essential to prove the seriousness of the Lean effort.

For labor costs, use the true value (not the raw cost) of labor. Your accounting department might have a fully-burdened charge rate, so use that. If there is no established charge rate, consider working one out with your coach, or get some help from accounting. Keep in mind the discussion about phantom savings. Small changes in labor don't produce any bookable (real) savings but the activity accomplished in the time that was saved might show some benefit.

True Cost of Labor (this is the cost to use to calculate how much to charge for in-house labor to do a Lean project)

All Return on Investment calculations should be based on the true cost of an hour of maintenance labor. The true cost of labor includes direct wages, plus a percentage factor for overtime, benefits, and indirect costs.

Benefits include the costs of health insurance, FICA (in the USA, the employer's contribution to Social Security), pensions, life/and disability insurance, and any paid perquisites. Indirect costs include indirect salaries (costs of all support people who don't show up on Repair documents, materials, supplies (uniforms, bulk materials, soap, etc.), costs of the shop (utilities, depreciation of facility, tools, insurance, and taxes), allocation of costs of corporate support, costs of money, and hidden or other indirect costs.

For example (cost per labor hour):
Direct hourly wages: $25.00

Chapter 29

Benefits at 25% of wage $ 6.25 (this amount does not include vacation and sick pay, which is covered in the next section)

Indirect costs: $ 9.00 (add up the costs of running the maintenance department and divide by the number of Work Order hours)

Total cost per hour $40.25

The total cost per hour must be increased to cover time that is paid for but does not appear on Repair documents. Consider your operation: of the 2080 hours straight time (52 weeks x 40 hours/week) available per year. Typical calculations show:

Annual Hours: 2080
less: vacation 160 hours (4 weeks)
           holiday 64 hours (8 paid holidays)
           paid sick leave 40 hours (5 sick days)
           All Other: jury, guard, training, union 32 hours
           (4 days per year for all other categories)
Total Annual time available 1784

If we start with a $25/hour, direct wage:
The actual burdened cost of time = $40.25 x 2080/1784 = $46.93
(In this shop example we might round up to $50.00 per hour)

Lean projects are a team production

Assign team members to specific aspects of the project. Some potential assignments might include: formal project write-up, data gathering, interviews, observations, taking readings, calling vendor engineering support departments, making presentation materials, giving presentations, formal write-up of findings, etc. Someone must be appointed as the point person accountable for delivery of the project. This person is the herder, to ensure that the project is completed, and the role can rotate through the team if the members stay together for several projects. The important message is that all members of the team have a role.

Once a Lean Project has been Chosen for Refinement

Planning

How complex is this project? If it is quite complex, then the project itself will have to be planned. If it is a simple project, only the work content will have to be planned (called micro-planning). The first step is to look at all aspects of the job. Some of the questions in the overall planning cannot be answered until the micro-planning is completed.

Use the following guide to write up and plan the successful project. Use the titles as guides, but an individual project might need some additional categories.

Lean Project Information Guide

Date: can be when you chose this project to plan.
Name of Project: Choose a cool name!
Description: Someone outside maintenance should be able to recognize what the project is about from this one- or two-line description. The more understandable, the better.
Team Members: Primary members. When the project is over, add a category listing support team members with everyone who helped even a little bit.
Objective: Goal of this effort in specific terms.
Hypothesis: What do you think will happen? Is there a theory you are testing?
Measures: What will you measure and how will you measure it?
What is the current situation? Always describe what the state of the area is now. Specific, measurable is better.
What savings are promised? Where will savings come from? Is this forecast phantom or real? How much do you estimate the savings will be?
What non-monetary savings are promised? Is anything else impacted, like comfort, speed of process, etc.
What process will be followed? What are you actually going to do? How are you going to do it?

| How much Labor is needed? This would be the labor to accomplish the project. Use your judgment about whether to include the time spent discussing the project and writing up the reports. |
| What parts are needed, and are they in stock? Create a Bill of Material for the project. Whenever possible, use stock items. Even better, use up any materials left from the last project. |
| How much total money is needed? Total cost of the project. |
| What other resources are needed? Do you need experts, NDT, other support? |
| Other: |

Lean Project Information Guide

Date:	
Name of Project	
Description	
Team Members	
Objective	
Hypothesis	
Measures	
What is the current situation?	
What savings are promised?	
What non-monetary savings are promised?	
What Process will be followed?	
How much Labor is needed?	
What parts are needed, are they in-stock?	

Once a Lean Project has been Chosen for Refinement

How much total money is needed?	
What other resources are needed?	
Other?	

Once the overall Lean Project is planned, it is time to prepare the "sales document" or Project Proposal so that you can pitch it to the administrator of the Lean teams or to management (if this is needed). It is essential to gain mastery of this process and to go through these steps for at least the first few projects so that the team can grasp all the steps, and to provide documentation of the process for future Lean Teams. Before you complete the process it would be wise to plan the work content of the job, to be sure your estimates are reasonable.

Proposal

The Proposal will include information from the Lean Project Worksheet. The worksheet will be transformed into a short narrative with tables, charts, and photographs as needed to tell the story. The proposal should answer all the questions of the Lean Project Worksheet.

The proposal should always start with an executive summary. This summary should sum up the project and the savings proposed, should occupy no more than half a page, and be titled Executive Summary.

You may be asked to present your project verbally. The presentation should be practiced (the verbal part) and reviewed in detail with your coach before turning it in, or presenting it to management. A good coach will test the presenter by firing questions or challenging the assumptions, and seeing how the project presenter reacts. This process is very useful training to make sure that the presenter feels comfortable with the material.

The written proposal (and the hand-out for the verbal proposal) should be neatly prepared (spell check it!). Solicit help from accounting and other groups to determine such things as the costs of labor, and overheads such as the amounts paid for electricity, etc. It is important to give credit where credit is due if you received help from outside the team. With some presentations it would be to your advantage if a representative from that outside group presents that section of the report (it gains credibility).

Use the language, formulas, and models already in use in the organization whenever possible. Also, use copies of existing reports whenever possible, such as downtime and production reports, which will make your Lean Proposal seem more familiar. Use copies of the source documents (usually in the appendix) as much as possible; such as actual bills, recorder charts, and measurements. If the Lean effort has a web site on the company Intranet, then some of the source information can be placed there for review.

CHAPTER

30

Micro-Planning

Begin the process of micro planning (usually called Job Planning) for your project. Planning Lean Projects is similar to planning maintenance jobs. Planning is the process of listing and assessing the elements of a successful maintenance job. Planning is not ordering or moving anything around in the real world, just listing items. You do not place orders or gather materials until the go-ahead is given and the project is put on the schedule.

The resources include labor, parts, tools, access to asset(s), permission, drawings, and information, and any other elements necessary to complete the project. If the project is a study or investigation, then decide who you need to talk to, and what internal or external resources will be needed. Planning also includes thinking out in detail how you're going to do the project (job steps), and what (if any) engineering help is needed.

Lean Project Planning check list (adapted from Planning, Scheduling and Coordination, a 2-day class presented by the author):

✓ Choose a Lean Project.

✓ Visit the Lean project job site and have a discussion with the team or customer (if needed).

✓ Find out whether this job, or a substantially similar job, has been done before (if so, look at the plan and see if it, or elements of it, can be re-used).

✓ Get approvals based on a Proposal containing a preliminary estimate of costs.

✓ Break the job down to its component tasks. List the job steps. Take the physical location and spaces around the equipment into consideration.

Chapter 30

✓ Evaluate each step for hazards.

✓ Analyze the job for risk to duration, scope, and process. Evaluate how the job might impact the whole company, line, train, or whole department, if it fails to work.

✓ Include ideas about whom (by skill set) and how many people can work efficiently in the space available. Establish the duration and work force needs that will be required to perform the work. Determine if there are any special licensing requirements.

✓ Determine if outside skills are needed, including such things as consulting, engineering, NDT, equipment operation.

✓ Clarify the sequence of skills required throughout the job to estimate the needs for coordination between crafts or other groups.

✓ Identify what's needed in terms of spare parts, materials, consumables, special tools, PPE (personal protective equipment), and equipment necessary to do the job. Determine if you have the items in stock (same answer as the project plan).

✓ Determine what essential reference materials are needed. This list might include drawings, wiring diagrams, and other Reference Documents.

✓ Planning is not completed until you make a list of everyone who needs to know what is going to happen.

✓ List all permits and clearances.

✓ Identify and list all parties that should be notified, and all processes that have to be re-routed, shut down, or backed-up.

✓ Preparatory and Restart Activities **must be** listed and coordinated, regardless of whether the responsibility for implementation belongs to operations, maintenance, or other department).

Project Micro-Planning Sheet (use this list as a guide to Micro-plan the Lean Project)

Labor (skills, licenses, hours, contractors, etc):
Parts, Materials, supplies, in stock or to be ordered:
Tools, equipment, to be rented or in-house:
Access to unit, any special notifications needed:

Micro-Planning

Clearances, permits, statutory permissions, lock-outs, other permission:
Engineering (formal or informal), drawings or sketches, wiring diagrams, other detailed Information:

Complete Planned Project Package for Lean Project (also adapted from Planning, Scheduling and Coordination, 2-day class presented by the author)

These steps will create a planned job package, which might have some or all of the following elements. Only the largest and most complex projects would need all these items. A small project might be only a work order covering the tools and materials.

Work order
Bill of Material. List all materials needed for the job, including an acquisition plan for major items. Determine whether the material is authorized inventory or directly purchased items.
Job plan with details by task, with step-by-step procedures. Estimated time for each step (task), summarized by resource group (trade) and for the total job.
If the job is complex, consider the use of the GANTT bar chart or CPM network chart to help plan the sequence of tasks and coordination of different crafts and crews.
Tool List, equipment list, rented equipment.
Requisitions for materials or tools (do not submit until project is scheduled).
JSA (Job Safety Analysis) or HACCP (Hazard Analysis and Critical Control Points) and Listing of Personal Protective Equipment.
A copy of all required permits, clearances and tag outs.
Prints, sketches, Polaroid/digital pictures, special procedures, specifications, sizes, tolerances and other references that the assigned crew is likely to need.

CHAPTER

31

Execution

Experts in Lean see waste, and make a few simple changes to help the organization get Leaner. That is one approach and a great way to get started. But to make it stick, the workers have to get involved, and management has to build reporting structures into its systems that will check up on the Lean effort. What this means is that the more people given useful roles, the better.

Execution of Lean Projects has a long and illustrious history. Just think of how much our ancestors did with their limited resources. This history lends itself to heroic feats of improvisation, and working behind the scenes to obtain needed resources. A great Lean team will think outside the box. However, be advised that thinking outside the box is sometimes threatening to the rest of the workers. This concern is especially true when those workers are not part of the process.

There was a famous group at Lockheed Martin, in Burbank, California, starting in the late 1930s, who called themselves the Skunk Works. The group developed of secret aircraft projects and could design and build an aircraft in record time. Although they were not a Lean team, they were the forefathers of the can-do attitude that is needed to be successful in Lean projects. The term skunk works today is also used analogously in other fields to describe any self-contained, semi-autonomous work-group or committee that directly manages its own projects. These teams were famous for 'borrowing' resources and people from other programs and re-purposing assets to suit their needs.

Lean teams will use resources that were left over from other projects; they will re-purpose resources, borrow expertise from other groups (informally), and above all, try things rather than endlessly analyze them. After the Lean process gets bedded down, some people will look forward to working with the Lean folks (or agitate to become one of them) because it is fun and uses more of their capabilities.

Your proposal documents are the blueprint for action in execution of the project and are to be used in the proposal as an outline of the types of data you will need to make your case. You want the final report (whether the results are positive or negative) to make an active statement (either positive or negative) about the hypothesis (what you think will happen).

Whatever you mention in the proposal, make sure you collect information during execution to write what actually happened with that topic.

- Before starting any experiments, read requirements for both verbal presentations and written reports. This preparation will ensure that you don't get two weeks into a project before you realize that you didn't collect some critical piece of data.
- Based on the decisions made in the proposal about metrics, set up the baseline measurements for the project. Sometimes the baseline can be obtained from existing records (all the better) stored in purchasing, operations, engineering, or maintenance departments.
- Consider keeping a Lean diary of the project as it develops, as scientists keep track of experiments.
- A resourceful Lean Project team will be on the look out for resources wherever they can find them, and they especially like free resources. Call vendors for additional help, or for new products to solve old problems. Be sure to communicate with purchasing about what any vendor is doing for the company.
- The coaches are there for ideas to be bounced off, and they will challenge you to increase the rigor of the whole process. Contact your coaches if you need resources.
- Ask yourself if your measurement methods are sound and complete. Is there anything that could account for the result that has not been taken into account?

Chapter 31

- Record the data that your experiment calls for. As results start to come in, be very careful to record everything accurately, even if it doesn't seem to apply at first glance.
- Ask yourself the question that is often asked in science; are the results repeatable by another team at another time?
- What happens if you are proved wrong? Even negative results are important. Do not warp the data to suit your ideas. Such distortion is a big no-no in science and in Lean Projects.
- Start to design the Project Report and Presentation to announce your results.
- One aspect of the announcement is to take a census of the size or kind of equipment to which your experiment applies. For example, in a project in a county to save money in pool chemicals, the census would be the number of pools.

Once execution is done and the results are collected, the team will discuss the results and decide what to say. The project proposal is now to be expanded into a Project Report and the Verbal Proposal is to be expanded to a Verbal Report of Results.

The Project Report and Presentation of Results should be constructed in such a way that a non-maintenance person can understand what you planned to do, what you actually did, how you planned to do it and how you actually did it, how and where the savings will come from, and finally, how much savings did you actually realize.

Putting the Finishing Touches on the Presentation

The goal of verbal presentations is to present your project in a way that will educate the other people in the room about the subject. The presentation should open up the listener to new knowledge and tell how you acquired that knowledge. Interesting stories (whittled down to their bare bones), clear photographs of the process, (short) videos of the process, and best of all physical parts or components that help tell your story, are always welcome.

One of the great basic presentation courses had some simple ideas to make a difference in the effectiveness of a presentation. Follow the four simple rules in the class "Presentation by Objectives" from Xerox Corp:

Think about who you will be presenting to.
How is the presentation important to them?
What do you think the bosses want from this process?
What would you like to get from this process?

Some ideas for the verbal presentation:

There are many excellent texts on giving successful presentations. Get one or two books (or take a class like the <u>Presentations by Objective</u> mentioned above) to help you sharpen up the presentation. Generally, the Power Point computer program may be used to guide the effort and give the attendees something to look at. Make hand outs of the important points, and of any complex areas. Hand outs are not necessary for everything (in a Lean environment we print as little as possible).

Chapter 32

In the wrong hands, Power Point can be extremely dull or, worse yet, annoying. If you look up Power Point rules you will find hundreds of relevant sites containing the hard—learned rules of good presentations. Some simple ones include:

➢ Don't give PowerPoint center stage. The project must have center stage.
➢ Make sure there is a logical flow to the slides.
➢ Each bullet is to be followed by a capital letter, and to have eight words or less.
➢ Keep the font style simple and do not use ALL CAPS, because they are harder to read.
➢ Use one simple transition for all slides.
➢ Lettering on all slides must be readable from the back of the room (check this explicitly).
➢ Avoid excessive details.
➢ **Remember, less is more**

People love pictures, and today with YouTube we all love videos. Use photographs, videos, graphs, catalog sheets, specification sheets (with captions), physical components or parts as appropriate, to support conclusion. But use some discretion with the visual elements. Make the images relevant to the topic and interesting. We are sophisticated consumers of visual images and amateurish attempts might work against your main message. We don't want people laughing at the videos unless that is a useful response to get our point across.

Lean is a team activity. Use several members of your team to present the findings. Assign point people (experts) to various parts of the project. Some areas to divide up the presentation are: what used to happen, what we did, what the results were, economics, and implications for the future. At the least, have everyone near the front of the room.

Solicit help from accounting, billing and other groups to determine the costs of labor, overheads, and amounts paid for electricity, etc. Give credit where credit is due if you received help from outside the team (getting such help is okay!). Lean is a company exercise, so the more people who are involved, and the more departments involved, the merrier. By adding outsider partners you also build a constituency for the project and the methodology.

Putting the Finishing Touches on the Presentation

Good managers are suspicious, or at least skeptical. Any numbers or conclusions will be looked at as unproven unless you can link your conclusions to existing, widely-accepted sources of information. To facilitate belief, use the following sources first:

- Use copies of existing reports whenever possible, including downtime and production reports.
- Use source documents as much as possible, such as actual bills, recorder charts, and measurements.
- Use language, formulas, and models that are already in use in the organization whenever possible.

Format of a verbal Lean Project presentation (after the project is complete)

Name of Lean Project (remember the importance of cool names)
Executive Summary with relevant dates
Names of all team members including the management coach
List specific, understandable, recommendations.
Introduce the elements of the project (high probability of success, low cost, short payback of investment, tools and materials and other resources that are available, good ROI, and low risk). Answer the question: Why this project was done?
What were the conditions or measurements before the Lean Project?
What you actually did
What happened?
What is the census of situations of this type?
How did the economics work out for the Project, and what would happen if the project was extended for the whole population?

You can liven up the presentation with a little story-telling concerning how you began, how you picked this exact project, any interesting stories about the project, was it fun? To make the presentation more interesting, weave the answers to the following questions into your presentation.

Chapter 32

Sample topic areas for stories to fill in the verbal presentation

General
1. How did the project begin?
2. Where did the idea come from (which list was it from, if any)?
3. Why did the project appeal to the group?
4. What projects did you discard? Why did you discard them?
5. What did you hope to prove? What did you prove?

The Project itself
6. Explain (in some detail—but remember your audience, what do they need to know?), what you ended up doing.
7. How did the project change over its life?
8. What steps did you go through?
9. How did you plan to measure your results?
10. What was required in materials, costs, and hours?
11. What outside resources were needed?
12. What were the results?
13. How did you end up measuring the results?
14. Was there a savings of any resource?
15. Was anything else impacted by your project (unintended consequences, both good and bad)?
16. (In your estimation) how easy would it be to ramp up your findings from a pilot (like the one in your project) to plant wide? Or company wide? Do you see any problems?
17. What would be the impact on specific benchmarks (choose the ones that apply, if any). Backlog of work available to complete, Emergency hours, Number of service calls?

Conclusions and actions to be taken
18. What is the best way to notify the right people of the new materials, techniques, tools, etc. that your project demonstrated?
19. What are your team's specific recommendations?
20. Do you have any ideas to publicize the project?

Project experience
21. What was your experience in doing the project?
22. Do you think that you picked the best project for your team?

Putting the Finishing Touches on the Presentation

23. How was it different to work on this type of project?
24. Did you enjoy the experience?
25. Do you have any recommendations to improve the process (please be specific)?

What do you say when the project fails?

The goal of Lean Projects is to reduce the inputs into making a product, providing a service, or increasing output. Sometimes, even with the best efforts, the project goes sour. The outcome for the presentation is somewhat different. The outcome of your presentation should increase your knowledge and the knowledge of the entire group, even if the results were negative.

Weave the answers to the questions that apply below, into your presentations. Valuable lessons are learned when a project goes down the tubes. Sometimes the most valuable information is gathered from failures and not from successes. We want to learn what happened. We want the details of what went wrong so that we all can learn the lessons. This might be a presentation to the Lean Team, rather than to outsiders to the team.

Some questions to address (in addition to others in the section above).
• Explain what you actually ended up doing?
• Where did the breakdown occur (here, a breakdown is when something went wrong with the project or there were negative results)?
• How did the team deal with the breakdown?
• How did the breakdown occur?
• When did the breakdown occur (exactly what was going on when the group stopped working on the project)?
• What might have avoided the breakdown in the first place?

Goal of written report: To present the history of your project in a step-by-step manner from the idea to the results. The report should be a travel log of the journey with facts and dates, successes and failures. The written report is primarily a project information document to help evaluate the results of doing the project. It should contain all relevant data (if practical) The report is also a teaching document. Be sure all names of team members, coach, and outsiders that provided any help, are in your written presentation.

CHAPTER

33

Publishing Lean Projects

When does anyone think about maintenance and its contribution to the success of the company? When do people talk about maintenance? If you were a fly on the wall in the lunch room, when would maintenance come up in people's conversations? How often is maintenance blamed for breakdowns, or for lack of responsiveness around breakdowns?

People think about maintenance when something breaks. People think about maintenance when there's a problem. People also tend to blame the messenger. Maintenance frequently has bad news for operations and others.

Can you imagine the operations people telling each other that they are so blessed with this maintenance department? That they feel that the maintenance department and the maintenance people are wonderful. You don't hear that much, though it could happen.

Until now I would bet you that when you did something to save the company money or effort, or when you eliminated waste, you didn't blow your own horn. I'd bet that, probably outside of three or four people close to you, no one knew that you even did anything.

I know this statement is true because I give personality tests to maintenance people. And almost half of the maintenance people, about twice as many as you would normally find, are what we call introverted in their personality profile. Maintenance people as a group don't brag a lot, and they're not particularly good promoters. Maintenance does not attract the kind of people that yell to the world, "Yes, we did this! Let's tell everybody!"

The result is that maintenance is undervalued and misunderstood, within the corporate hierarchy. Corporations can make decisions that are not good maintenance decisions with impunity, because they haven't heard from the maintenance folks. They haven't heard because more than half the maintenance folks have introverted personalities.

Whenever we eliminate some bit of waste, we need to know what was being used before, what's being used now, and what's the difference between the two. With a scientific approach, any controlled experiment will provide this data. It won't be news to people reading these words, but a lot of people in maintenance will sneak in on a Saturday, do something to improve things, and then go home and tell no one. And the action could have made a major difference in the way things were happening at that point. So collect reports, prove the results, and tell people about it.

I want to go back to a story from the beginning of this book. It is a Star Trek story about the chief engineer of the Voyager, B'Elanna Troy. Her mission was simply to get more power from the Warp engines. Her role is like the future of maintenance. When maintenance takes on Lean, it moves toward its own future.

We become the agents of producing products with less input. Our expertise is essential in the next generation of production. When we start measuring waste, and we start doing projects in waste removal, waste reduction, or waste elimination, we shift into the future of maintenance. By the way, this movement is also the future of all commerce.

For maintenance to be viewed differently, we have to view ourselves differently. The way we start is to change the stories that circulate in the maintenance department. Once the Lean program is running, after every Lean Project, you need to create and tell the story of how the maintenance department saved this money or that energy.

The next step is to start publishing the stories of these saves for others. We want to get people to think about maintenance in other ways than just as people to call when there's a problem, or people to blame when something doesn't happen right. Start publishing and talking about these kinds of successes—when we find air leaks, fix the steam system to increase efficiency, reduce downtime, and so on.

Chapter 33

Some ideas for the written report:

- Neatly prepared and complete. The report should be made with a finish and production values similar to those of any other presentation in the company.
- Can follow the logic of the verbal presentation, or can be chronological.
- Use graphics, visuals, pictures if possible. Exhibits from the verbal presentation can be used, but remember, there will be no one there to explain the charts, graphs and pictures, so more words may be needed.
- Should be reviewed by the coach before being turned in. Another good idea is have someone who is not from the area read it, to see if they can follow the logic and the arguments.
- It is vitally important to get everyone involved. Solicit help from accounting, billing, and other groups, to determine the costs of labor, overheads, and amounts paid for energy (electricity, etc.). Give credit where credit is due if you received help from outside the team.
- These same rules apply to information. Use copies of existing reports whenever possible, such as downtime and production reports. Use source documents as much as possible, including actual bills, recorder charts, and measurements. And finally be sure to use the language, formulas, and models already in use in the organization whenever possible.

Format of Written Presentation: is very similar to the format of the verbal presentation, but is just less skeletal and more fleshed out. The written report relies less on the support of the presenter. Some guidelines for the written report:

Publishing Lean Projects

Name of Lean Project (remember the importance of cool names)
Executive Summary with relevant dates
Names of all team members including the management coach and others that contributed to the project.
Table of contents
Specific recommendations.
Introduction to the six elements of the project (high probability of success, low cost, short payback of investment, tools, materials, and other resources that are available, good ROI, and low risk). Why was this project chosen?
Project planning sheet
What were the conditions or measurements before the Lean Project?
What you actually did
What happened?
Project investment analysis sheet
What is the census of situations of this type?
How did the economics work out for the Project, and how would they work if it was extended for the whole population?

34

How a School District could Save $1,000,000 in O & M Costs and Improve Service!

Summary

A school district in the Southeast embarked on a Lean Maintenance effort. Their effort was unique in that the maintenance tradespeople, grounds-keepers, support staff, and supervisors, were the key participants. Over 135 personnel were trained in cost reduction techniques. The teams were told to choose their own cost reduction projects, and were given three weeks to get results. Members of these teams had an average of a high school education, and many did not even have that. Several teams had only a few English speakers.

Methodology

We met with the entire maintenance work force in groups of about 30. At the first meeting, Mr. Levitt taught the concepts of cost avoidance, life cycle costing, and calculating return on investment. He set up multi-skilled teams. Supervisor and manager coaches were assigned to each team to help with logistics and resources. Through a structured process, each team first brainstormed a list of possible projects, then chose a cost-reduction project that appealed to them.

The teams worked on their own for three weeks. In a subsequent meeting, Mr. Levitt taught the teams about both written and oral presentation skills. In the finale, each of the 28 teams presented a

How a School District could Save $1,000,000 in O & M Costs and Improve Service!

written and an oral report about their project. Other class members were instructed on how to ask questions to determine if the savings were real. Some of the projects were studies, and were not actually carried out, but others were fully executed.

The Projects

The 28 projects listed below are summaries of the reports presented by the teams. When and if all these projects were implemented, they represented savings to the district of over $1,000,000 over the next few years.

Gold ★ **Effect on electricity consumption of increasing the air-conditioning set point of a school from 75.5° to 77° F.** The school-wide set point in the J school was raised from 75.5° to 77°. No complaints were registered. Electricity usage dropped by 10% resulting in a $7000 a year savings per school, with an immediate total approaching $315,000 (45 schools on DDC controls) annually for the district. Additional recommendations were made for savings from installation of new thermostats in modular units. A Gold star was awarded because the project needed little investment but it generated large potential savings, and immediate returns. The savings in the schools with DDC (Direct Digital Control), could help fund projects to install these modern controls at the rest of the schools in the district.

Silver ★ **Impact of the use of stabilizer on the consumption of chlorine in swimming pools.** The team added $225 worth of stabilizer to a test pool. They charted chlorine usage for one week before and one week after stabilizer was added, and chlorine usage in the test pool dropped by $80 per week. Potential savings was $4000 per year per pool. A Silver star was awarded because investment is small, and payback is large and immediate. This technique could easily be applied to all district pools where stabilizer is not being used. The district in question operates 50 pools and could generate $200,000 of savings per year.

Bronze ★ **mpact of an Ever-pure filter system to prevent corrosion failure in kitchen boilers.** The manufacturer will increase the warranty from 1 to 7 years if an Ever-pure filter is installed in the boiler system. Installation parts and labor are about $550. Current failure rates of 1 liner every 2 years will result in a savings of about $2684 in 7 years for each unit equipped. Fourteen (14) proposed Viking units would yield $37,500 in avoided maintenance costs over the 7-year warranty period. In addition, the school system contains more than 50 boilers of other makes and models. with equally- high failure rates. Use of the filters might help generate additional savings of $134,200. This project was chosen because of the ease of the savings (from a warranty), the real reductions in labor, and the high cost of parts. If the technology proves to be effective in a pilot experiment, the project can be extended to all the boilers in the school district.

Comparisons between circulating pumps. This project compares an existing cast iron circulating pump with a smaller and lighter, stainless steel pump. The smaller pump has now been in use in the district for 7 years. Analysis of replacement cost, reliability, energy usage, and complexity of installation, shows that the smaller, stainless steel pump is clearly superior. The stainless steel pump costs $150 less to purchase, is more reliable, and uses about $7/yr less electricity. The recommendation is that all circulating pumps be replaced with stainless steel pumps as they fail.

Impact of re-lamping and cleaning fixtures on light output and electrical usage. In one classroom where there were lighting complaints (room 12, Croton), the team replaced the tubes (with energy-efficient ones), and ballasts (with magnetic ones), and cleaned the fixtures. The candle power measured at desk level increased from 12.5 to 56.8 (the specified standard is 60). The amperage consumption at the breaker went down from 14 to 8. The total cost was $305, including labor at $25/hr. Electricity savings was $220/year. Several ballasts were found to be leaking and were hazardous. The Recommendation was to create an annual campaign in which each school was to choose its worst-lighted 2 or 3 rooms. Teams would then re-lamp, re-ballast, and clean the fixtures. Based on the first project for re-lamping, installation of high-efficiency ballasts, and utility rebates, the school system should look into similar changes elsewhere.

Comparisons of individual pothole repairs to parking lot resurfacing. Calculations show that resurfacing of parking lots is economic if there is more then one pothole per 1000 square feet. Pothole repairs are more costly, last only a short time, and are usually done on overtime (to minimize interference to the schools). Resurfacing lasts an average of 6 years and enhances the look of the school.

Comparison of direct-drive exhaust fans to belt-driven fans. A study shows that belt- driven fans are more susceptible to maintenance calls and are also noisier than direct-drive fans.

The recommendation is to purchse direct-drive fans for new installations. Direct-drive fans are less expensive to purchase, quieter in use, and require lower maintenance over the life of the unit. Replacing existing units is not indicated unless extensive other work is needed.

Two studies were done on doors:

Comparison of the costs of the three dominant door systems used on the beach side of the district, proved that the best was the aluminum door with the heavy duty aluminum jamb. It was calculated thet this combination would save over $2000 compared with wood and over $800 compared with steel, per opening, over the 15-year life of the door. The recommendation was that, whenever major door work is undertaken, doors should be replaced with aluminum, and that aluminum be specified on all new construction.

Identification of schools with the fewest door service calls. The project team found that the schools with aluminum frame doors had the fewest service calls by an order of magnitude (in agreement with the findings of the project above). The team also found that door installation is a major factor in maintenance. Doors that can open all the way, up to 180 degrees, or to a wall with a magnetic or friction catch, have the fewest problems. The recommendation is that aluminum-framed doors be used with careful installation for fewest calls.

Chapter 34

There were two projects on safety of playground surfaces.

Sand versus loose pack recycled tires for safety surfaces. The team studied safety surfaces from the points of view of cost and safety. Sand needs to be replaced at least once a year, carries some diseases, and does not provide an effective safety surface because it compacts when wet. Surfaces made from recycled tires cost about twice as much as sand, but form a more effective safety surface (a 6-inch thickness will break a fall from a height of 7 feet), and don't support or attract as many pests as does sand. The rubber also carries a 24-year warranty. If the rubber is not dispersed it will provide a return (ROI = about 50%) starting in the second year. There is also significant (but uncalculated) savings from avoiding the effects of sand on the wood floors when the sand is carried on shoes, into the school).

Safety surface comparison between sand, mulch, shredded recycled tires, and a poured-in-place recycled tire system. The conclusion was that the poured-in-place system was by far the best, but the most expensive. Although material costs are only $14/sq foot including site work, overall costs and upkeep would drop with the poured-in-place system. Poured-in-place systems are the safest, with 3 inches being the recommended depth for protection from a 9-foot fall. Savings would come from reduced sand costs, improved health and safety, and complete ADA access. For both types of project, the state provides grant money of up to $150,000 for districts that use re-cycled tires. These moneys would pay for over 10,000 square feet of playground surface a year. This grant would almost cover the district's needs if the loose system was used. Because of the significant safety improvement and the handicapped access, this project might be funded by PTA fund raising.

Energy usage and comfort were the subjects of four projects.

Effects of window film applied to a non-air-conditioned room facing east. It was found that the window film reduced the ambient temperature by an average of 5° F. A letter of thanks was sent by the teacher. The investment was $205 for the film and $1200 for labor at $25/hr.

How a School District could Save $1,000,000 in O & M Costs and Improve Service!

Investigate energy usage and comfort of classrooms with slat windows versus single—hung windows. The team replaced six windows in a modular classroom with single-hung windows (that were recycled from a prior new-window installation). With the same air conditioning load, the room temperature dropped 5° F, and humidity was reduced by 4% (reducing the probability of mold formation). The investment costs included minimal material, but $600 worth of labor at $25/hour. The recommendation was to replace any slat windows with single-hung units where the recycled windows will fit, starting with the classrooms that are hottest in summer.

Impact of reducing humidity on room-cleaning costs. Even if dehumidifiers were used (the least efficient way to reduce humidity), there is a monthly savings of custodial labor (less the cost of electricity) of over $50. This study proves the need for lower humidity where mold is a problem.

Test to measure the impact of cleaning heat exchangers in air conditioning units. Dirty heat exchangers are not only inefficient but they negatively impact air quality. After a 4-hour cleaning (costing $100), tests demonstrated an 8% increase in air flow and a 7% decrease in humidity, with no increase in electricity usage.

Projects covered many other different topics, and here are 3 different areas.

Medium-quality versus high-quality paint brush comparisons. This project showed that better quality brushes hold paint better, cover surfaces without streaking, and can sometimes cover an area with one coat. These attributes led to the recommendation of the study that time can be saved and quality improved by allowing full-time painters to use better brushes. Estimated savings were about $500 per painter, per year. It was further recommended that occasional painters continue to use the lower-quality brushes.

Impact of re-lamping using T8 bulbs and electronic ballasts on consumption of electricity. This study of specification sheets shows a calculated savings of $325 in electricity per classroom year.

Chapter 34

Paper recycling drive. The office staff started a drive to recycle usable paper into scratch pads. The team also arranged for a local charity to pick up recyclable materials, proving that over $600 per year can be saved from the office budget.

Temporary fire alarms. In several schools, the inability to get fire alarm parts for old and obsolete systems required that the school operate without a working fire alarm until parts arrived. This team designed and built a movable, manual, back-up notification system at a complete cost of less than $500. No savings were indicated, but there was significant reduction in danger and liability.

Chlorine usage analysis. The team looked at the current procedures that allow the chlorine service company to come and go without challenge, and add chlorine to the tanks without any district verification. Recommendations include spot checking of deliveries with gauges, having the contractor sign in, and considering installation of flow metering equipment.

New equipment purchase considerations.
Comparisons between a tractor/bat-wing cutting combination versus a specialized mower. The distributor arranged for a side-by-side cut test. The results showed that the new mower was more maneuverable but had a narrower cut, though the effective cut speeds were the same. The new mower was much safer for the operator to transport, and for anyone who was behind the cutters.

Comparisons between wood and aluminum used in steps and landings for modular classrooms. Aluminum steps cost almost double and last four times longer than wood steps. Aluminum steps are also easier to move, do not require annual maintenance, are safer, easier to adjust, reduce district labor requirements for carpenter labor, and can be adapted to different requirements more easily. Savings could run to $180,680 over 20 years in this application.

How a School District could Save $1,000,000 in O & M Costs and Improve Service!

Three projects proved the correctness of prior district decisions.

New looks at using washable dishes versus plastics disposable dishes. The district took out their dishwashers some years ago. This project revisited the issue. The conclusion was that the current decision is still the best one. The group did recommend considering recyclable paper products where possible.

Testing premium edger blades versus the currently-used inexpensive 10-inch blades. Premium blades cost $8 each and the current blades cost $2 each. The blades were tested under real conditions during alternate weeks. The conclusion was that it would be cheaper to stay with the blades currently used by the district.

Comparison of existing roof cement with a new product. Tests showed that the currently-used product, Tamco Roof Cement, is superior to the new product introduced by Sunniland.

Several projects are ongoing and will not have results for a year or more.

PM versus non-PM. This study looked at the frequency of quick calls in buildings with an active PM program, and demonstrated that quick calls were reduced with PM. Scheduled maintenance was shown to increase for a year and then drop off. Continued study is recommended.

Comparisons between sheet flooring and vinyl tile. Flooring and floor care are a big expense in the school district. The installed price of sheet goods is higher than for tile. The experiment would compare the two materials in use in a classroom. Armstrong (the manufacturer of both products) will donate the materials for the test. This is an ongoing project with preliminary results expected in a year.

Comparisons of chemical and magnetic water treatment for cooling towers. Results of interviews with current and past users of both systems are inconclusive. The team recommends close monitoring of Central and Meadowlane schools (which have new magnetic installations) for costs and amounts of maintenance required.

Chapter 34

Total Cost

Presentation of sessions including all travel expenses, development of customized training materials, duplication costs, follow-up consultation to help define and prioritize projects, write-ups, other support	$40,000
Time in class 150 people x 8 hours @$25/hour	$30,000
Time on project outside class, estimated average of 2 hours per person. Total 300 hrs @ $25/hour	$ 7500
Materials for projects at an average estimated at $90/group	$ 2500
Management time to set up, coach groups, rewrite projects, review outcomes, and meetings, estimated at 200 hours @$50/hour	$10,000
Total cost	**$90,000**

Lean Maintenance Bibliography

A New Strategy For Continuous Improvement: 10 STEPS TO LOWER COSTS AND OPERATIONAL EXCELLENCE By Phillip Slater, published by Industrial Press. 2006

Complete Guide to Predictive and Preventive Maintenance By Joel Levitt, published by Industrial Press. 2003

Eaton Haughton of Econergy Engineering Services LTD Box 352, Ocho Rios, Jamaica. Phone (876) 974-5064 and E mail: econergy @infochan.com or visit his Web site www.easi-caribbean.com (Energy saving ideas)

Glossary was adapted from an article in Assembly Magazine, February 10, 2005 can be found at: http://www.assemblymag.com/CDA/Articles/Web_Exclusive/ca93 022c106c9010VgnVCM100000f932a8c0

Imants BVBA newsletter (a European consulting company) can be found at www.managementsupport.com

Lean Machines article by George Koenigsaecker, President, Lean Investments, LLC, Muscatine, IA gkkaizen@aol.com

Lean Maintenance by Ricky Smith and Bruce Hawkins, published by Elsevier NY. 2004

Lean Manufacturing Pocket Handbook by Kenneth Dalley, published by DW Publishing. 2003

Lean Solutions by James Womack and Daniel Jones, published by Free Press NY. 2005

Lean TPM, A Blueprint for Change by Dennis McCarthy and Nick Rich, published by Elsevier NY. 2004

Maintenance Planning, Scheduling & Coordination By Don Nyman and Joel Levitt, published by Industrial Press. 2001

Managing Factory Maintenance 2nd Ed. By Joel Levitt, published by Industrial Press. 2005

Managing Maintenance Shutdowns and Outages By Joel Levitt, published by Industrial Press. 2004

Lean Maintenance Bibliography

RCA RT (Root cause Analysis) GPO Box 407, Melbourne, VIC 3001 Australia 61-3-9697-1100 www.rcart.com.au

Ronald M. Schroder, speech on spare parts at the Reliabilityweb.com maintenance conference in Waikiki, Hawaii. 2007

Smart Inventory Solutions 7 Actions for MRO and Indirect Inventory Reduction By Phillip Slater,published by Industrial Press. 2006

The Kaizen Pocket Handbook by Kenneth Dalley, published by DW Publishing. 2003

The Machine That Changed The World, The Story of Lean Production, by James Womack, Daniel Jones and Daniel Roos, published by Harper Perennial. 1991

Lean Glossary

Cell: an arrangement of people, machines, materials, and equipment—with the processing steps placed right next to each other in sequential order—through which parts are processed in a continuous flow. The most common cell layout is a U shape.

Circle game: Draw a circle on the floor and observe an area or process with the intent of identifying waste.

Continuous flow: a concept in which items are processed and moved directly from one processing step to the next, one piece at a time. This concept is also referred to as "one piece flow" and "single piece flow."

Corrective Maintenance: Work on assets that have not yet failed, the need for which was identified by inspection.

Cycle time: the time required to complete one cycle of an operation. If cycle time for every operation in a complete process can be reduced to equal takt time, products can be made in single-piece flow. See "takt time."

Error proofing: a process used to prevent errors from occurring or to immediately point out a defect as soon as it occurs. If defects don't get passed down an assembly line, throughput and quality are improved. See "poka-yoke."

Flow: the progressive achievement of tasks along the value stream so that a product proceeds from design to launch, order to delivery, and raw materials into the hands of the customer, with no stoppages, scrap, or backflows.

Jidoka: one of the two pillars of the Toyota Production System. Jidoka provides machines and operators with the ability to detect when an abnormal condition has occurred and immediately stop work. Jidoka highlights the causes of problems when they first occur, which leads to improvements in built-in quality by eliminating the root causes of defects.

Lean Glossary

Just-in-time: a system for producing and delivering the right items at the right time in the right amounts. The key elements of just in time are flow, pull, standard work, and takt time.

Kaizen: a Japanese word that means "continuous improvement." Kaizen refers to small, incremental improvements in an activity, designed to create more value with less waste. A kaizen event is a highly-focused, action-oriented workshop that typically involves a team of five to 15 individuals, usually lasting three to five days. The goal of a kaizen event is to concentrate on improving one specific process. There are two types of Kaizen that apply to Lean Maintenance; Flow Kaizen—value stream improvement, and Point Kaizen—Waste elimination.

Kanban: a Japanese word that means "card" or "visible record." It refers to a small card attached to boxes of parts, that regulates pull by signaling to upstream production and delivery points.

Kitting: is putting all the parts and tools for a job into a box or tote called a kit. This procedure eliminates time-consuming trips from one parts bin, tool crib, or supply center to another to get the necessary material.

Lead time: the total time it takes to receive parts that were ordered. It is divided into internal lead time (time from when the part is needed—falls below the reorder point until the vendor gets the Purchase order) and external lead-time (time from when the vendor gets the PO to when the item is delivered on your receiving dock).

Lean Maintenance: A maintenance philosophy that reduces the input resources or increases the outputs in the maintenance function.

Lean manufacturing: a manufacturing philosophy that shortens the time between the customer order and the product build and shipment by eliminating sources of waste. By attacking and eliminating waste within a plant or process; Lean manufacturing reduces costs.

Lean Glossary

One-piece flow: the opposite of batch production. Instead of building many products and then holding them in a queue for the next step in the process, products are put through each step in the process one at a time, without interruption. This method improves quality and lowers costs.

Point of use: a technique that ensures that people have exactly what they need to do their jobs—the right work instructions, parts, tools, and equipment—where and when they need them.

Poka-yoke: a Japanese word that refers to a mistake-proofing device or procedure used to prevent a defect during the production process. See "error proofing."

Pull production: the opposite of push production. With Pull production, products are made only when the customer has requested or "pulled" them, and not before. Pull production, prevents the building of products that are not needed.

Right sizing: a process that challenges the complexity of equipment. The process examines how equipment fits into an overall vision for how work will flow through the factory. When possible, right sizing favors smaller, dedicated machines, rather than large, multipurpose, batch-processing machines.

5S: a lean tool used for workplace organization and standardization. Benefits include prompt problem detection and clear standards. In addition, routine disciplines are established to keep the workplace in order and ensure that materials are in the correct location to maximize productivity. The 5S's are sifting, sweeping, sorting, sanitizing, and sustaining.

Sanitizing: one of the 5S's. Sanitizing is the act of cleaning the work area. Dirt is often the root cause of premature equipment wear, safety problems, and defects.

Lean Glossary

Sifting: one of the 5S's. Sifting involves screening through unnecessary materials and simplifying the work environment. Sifting is separating the essential from the non-essential.

Six Sigma: a standard of operational excellence used in lean manufacturing environments. It is a process that designs and monitors every-day business activities in ways that minimize waste, while increasing customer satisfaction. Six Sigma objectives are directly and quantifiably connected to the objectives of the business.

Sorting: one of the 5S's. Sorting involves organizing essential materials. It allows the operator to find materials easily when needed because they are in the correct location.

Standardized work: a precise description of each work activity specifying time, the work sequence of specific tasks, and the minimum inventory of parts on hand needed to conduct the activity.

Standard work instructions: a lean tool that enables operators to observe the production process with an understanding of how assembly tasks are to be performed. The approach ensures that the quality level concept is understood, and serves as an excellent training aid. It enables absentee replacement individuals to easily adapt and perform the assembly operation.

Sustaining: one of the 5S's. Sustaining is the continuation of sifting, sweeping, sorting and sanitizing. It is the most important and the most difficult, because it addresses the need to perform the 5S's on an on-going and systematic basis.

Sweeping: one of the 5S's. Sweeping involves collecting non-essential items and removing them from the work area.

Takt time: a reference number that is used to help match the rate of production to the rate of sales, that is, the rate at which customers require finished units to be delivered. Takt time is determined by dividing the total available production time per shift by the customer demand rate per shift. "Takt" is a German word for pace or beat.

Toyota Production System (TPS): a system developed by Toyota Motor Corp. to provide the best quality, at the lowest cost, and with the shortest lead time, through the elimination of waste. TPS is often illustrated as a "house" comprised of two pillars: Just-in-time and Jidoka. TPS is maintained and improved through iterations of standardized work and kaizen.

Value: a capability provided to a customer at the right time and at an appropriate price, as defined by the customer.

Value stream: the specific activities required to design, order, and provide a product, from concept to launch, order to delivery, and raw materials into the hands of the customer. Whenever there is a product for a customer, there is a value stream. Up to 90 percent of the actions and 99.99 percent of the time along a typical value stream, can consume resources, but create no value for customers.

Value stream mapping: the process of directly observing the flows of information and materials as they occur, summarizing them visually, and then envisioning a future state with much better performance. The process raises consciousness of the enormous waste of time, effort, and movement that occurs. The relevant actions to be mapped consist of two flows: orders traveling upstream from the customer, and movement downstream from raw materials to the product delivered to the customer.

Visual controls: any devices, GANTT charts that help operators quickly and accurately gauge production status at a glance. Progress indicators and problem indicators help assemblers see when production is ahead, behind, or on schedule. The controls allow everyone to see the group's performance instantly, and increase the sense of ownership in the area.

Lean Glossary

Waste: anything that does not contribute to transforming a part to suit the customer's needs. There are seven types of manufacturing waste: production exceeding immediate demand; excess work in progress and finished goods inventories; scrap, repairs, and rejects; unnecessary motion; excessive processing; wait time; and unnecessary transportation.

Work in process (WIP): items between machines or equipment waiting to be processed.

This glossary is partially adapted from an article in Assembly Magazine February 10, 2005, *Lean Manufacturing Pocket Handbook* by Kenneth Dalley and other sources

Index

Index